Kombucha
The Miracle Fungus

by Harald W. Tietze

GATEWAY BOOKS, BATH, UK

This edition, April 1995
Published by GATEWAY BOOKS,
The Hollies, Wellow,
Bath, BA2 8QJ, U.K.

Revised reprint May 1995

Copyright © 1994 Harald W. Tietze

Distributed in the U.S.A. by
ATRIUM PUBLISHERS GROUP,
3356 Coffey Lane, Santa Rosa, CA 95403

First edition July 1994
Revised reprints Sept, Nov 1994, Feb, April 1995
published by: Tietze Publications,
Bermagui South. NSW, Australia

Cover design by the Design Studio, Bristol
Photograph courtesy Kevin Redpath

Text set in Times $10^1/_2$ on 13pt.
by Oak Press, Castleton
Printed and bound by
Redwood Books of Trowbridge

British Library Cataloguing-in Publication Data
A catalogue record for this book is
available from the British Library

ISBN 1-85860-029-4

CONTENTS

FOREWORD TO THE SECOND EDITION

Every morning we start the day with a refreshing tea-drink. Nothing unusual about that you might think, except that our tea is no ordinary brewed beverage. It is Kombucha, the revitalising and delicious drink which has deservedly earned itself the name of 'miracle fungus'.

Kombucha is an ancient, fungus-like living organism which, when brewed with sweetened tea, produces a wonderful, health-enhancing drink. We were introduced to this immune-boosting treatment through a cousin in California, the sole survivor in a family who had all contracted cancer through a polluted public water supply. Kombucha encourages sharing, for it creates an offspring every week, and one finds someone to give it to. We brought ours back to Britain in June 1994, and have since given away dozens to our publishing colleagues, authors, family and friends. Most have adopted it successfully, reporting definite and individual improvements to their wellbeing and health.

Its use quickly becomes a way of life, enabling us to do something practical at home for our good health and for our family, costing virtually nothing. We can then have the added pleasure of passing Kombucha on to our friends and neighbours, also for their good health, happiness and wellbeing. It is easy to understand why its production and consumption in rural Russia were regarded as a sacred ritual!

We have now heard that taking Kombucha is empowering the Aids community in California with its immune-enhancing effects, and its fame has rapidly grown as a marvellous health drink. The British media have noted its rising popularity in the UK. At a time when the amount of pollution in our environments has grown alarmingly, our immune systems are going to need all the help they can get. What could be better than a natural source of immune enhancement with a 2,500 year pedigree of healing - and no bad press at all!

This book has an interesting history. It was first published in Australia in 1994 and was circulated initially for the German speaking community there. Since then it has taken the Australian health market by storm, and the first edition in English to be published by Harald Tietze is now in its fifth printing.

We are pleased now to be publishing the book on the other side of the world and help in its wide dissemination. In this new edition there is more information on how to produce the Kombucha drink, with advice on varying the taste and therapeutic effect with fruit and herbs. Anecdotal accounts of how the Kombucha therapy has helped people mounts and surprises us daily. It would be misleading to call it a panacea for all ills, but it is interesting how each individual seems to receive some specific help, often in a way they didn't anticipate.

Alick and Mari Bartholomew, Wellow, March 1995.

INTRODUCTION

Kombucha is an ancient food and healing source originating in Asia. Thanks to a growing interest in Eastern healing methods, Kombucha travelled via Russia into Western Europe. In those countries, Kombucha was regarded as a pleasant health-giving beverage which contributed to a balanced diet. On its way to the Western World this potion has often appeared to have performed miracles. It acquired names such as magical fungus, miracle fungus and elixir of long life because of its success in the treatment of modern illnesses. While Kombucha was already a popular healing remedy in the 1920s and 1930s in Russia, Czechoslovakia, Germany and Austria, the fermented drink almost disappeared during and after World War II. This was because sugar and tea were in short supply.

In the 1960s Kombucha was brought into Germany and Austria from Eastern Europe. In the 1970s, when the standard of living improved, it became known again as a miracle beverage; the trials and medical research into its effects were conducted mostly in Germany and were published in the German language. Articles about it appeared in many newspapers.

In modern societies, with their abundance of sterilised, homogenised or preserved food, a microbiological digestive aid such as Kombucha can work wonders. Many doctors and naturopaths recommend Kombucha-drinking for all types of illness.

Kombucha, like many other products of natural medicine, should not be seen as a cure-all. There is no such thing. Health is the balance of body, emotions and mind. When reading reports of the success, for example, of large doses of vitamins in curing certain illnesses, the question arises of how long it will take for negative side effects to manifest themselves. In the absence of natural dietary balance there can be no health. When curing an illness one should not only treat the symptoms, but also the cause.

It is exciting news that Kombucha is gaining in popularity and is being recognised as a remarkable life-enhancer. This wonderful living organism produces a refreshing, revitalising and delicious drink, easily made at home by all the family. Among its main claims, Kombucha strengthening the immune system, so helping in a cure for many illnesses. It also detoxifies, cleansing the blood, and strengthening the kidneys. Kombucha also plays a vital role in regenerating bowel flora. The use of Kombucha as an aid to healing is right at the top of my list, to be used with all other forms of holistic medicine.

Without Kombucha

Turn the book upside down to see the results of drinking Kombucha!

Part I:
WHERE DOES KOMBUCHA COME FROM?

ORIGIN OF THE KOMBUCHA 'FUNGUS'

No one can say, for sure, how and where the Kombucha fungus originated, but we do know that it has been used for at least two thousand years. It does not really matter whether the Chinese, Koreans, Japanese or Russians were the first to ferment the fungus. To be exact, it isn't really a fungus, as such, but rather a community of yeast and bacteria. One authority describes it as a lichen. The origin of the name could be Japanese, with 'kombu' standing for the brown tea algae and 'cha' for tea. The best description, in my opinion, is that of the herbalist, Pastor Weidinger, whom I have used as a source throughout this book. He says:

"Kombucha' tea is an ancient East Asiatic beverage which came out of the ocean. For three years I was a missionary on the island of Taiwan. This South-eastern coastal region with its subtropical climate and extended growing period made ideal conditions for growing the tea, which was farmed in large areas. This province is regarded as the origin for 'tea', ie. in Latin 'Thea'. The original name given by the Chinese writer Kuo-Po to the beverage extracted from these leaves was 'Tu' or 'Tschuan'. Today it is called 'Ch'a'. In the province of Fuken, however it is still called 'T'e'. This has given me a closer understanding of the word tea. K'un-Pu-ch'a, a tea-like wine......"

"My missionary activities also required my travels to the islands of Quemoi and Matsu which are situated close to the mainland near the Province of Fukien. I was very impressed by a beverage which was served to me by the local people which had a sweet-sour taste and was very refreshing in the hot climate of the

area. Was it wine which tasted like a delicious tea or an unusual tea that tasted like a rare wine? Notably, after drinking this beverage, I not only felt stronger after the long and tiresome walk, but strangely enough, felt healthier. In particular it helped my metabolism a great deal in this climate and made me feel very relaxed. When I asked what it was, 'K'un-Pu-ch'a' came the reply. I was startled. 'Tea which came from life in the ocean'? Already in the Tsin-Dynasty, about 221 BC, it was known and honoured as a beverage with magical powers enabling people to live forever. The tea was given different names. One of the most famous was the 'Godly Tsche'. This particular tea was used as a remedy for chronic gastritis. People also tell of the Korean medicine man named Kom-bu who in the year 414 prescribed the tea to heal the Japanese Emperor's disorders. The 'Godly Tsche' came from China, via Korea to Japan where it was given the name 'Tsche of Kombu'."

KOMBUCHA IN THE WESTERN WORLD

The beverage eventually reached Europe via Mongolia and Russia. When I mentioned Kombucha in an article published in Australia dealing with traditional medicine, I received countless letters from people who originated in the old eastern parts of Germany and who now live in Australia. They recalled their grandmothers giving them a refreshing beverage from a large pot, on top of which floated a type of fungus. Others remembered the recipe being brought by Polish workers from the East. In World War I the fungus-tea disappeared because of the sugar shortage.

More reports about the existence of this magical fungus began to emerge in the 1920s, however, when researchers started to show an interest in its properties. Their studies were suspended with the beginning of World War II. Kombucha only really became popular again in the 1960s and 1970s, probably as an antidote to unhealthy eating habits. The Western media took up the topic. Reports came out of Kargasok in Russia of people who reached a great age. This appeared to be linked with the drinking of Kombucha, which is why it is also known as 'Kargasok Tea'. According to a report, an 80 year old woman gave birth to a healthy baby, fathered by a 130 year old man!

The most famous research results come from the University of Omsk in Russia and, in the West, from the research of Dr Rudolf Sklenar. Dr Sklenar's research is often mentioned in the German press. He came from Eastern Germany where Kombucha has been used since the turn of the century among ordinary people. He studied medicine in Prague and had his first contact with the Kombucha fungus in a monastery. He worked with the culture during the second World War and based his scientific work upon Kombucha. In the 1960s he published his research in the scientific and general press, which resulted in increased awareness of Kombucha's healing and strengthening properties. Dr Sklenar used Kombucha successfully for diabetes, high blood pressure, all types of digestive problems, stomach and bowel illnesses, rheumatism and gout.

Dr Sklenar's main area of work became the biological treatment of cancer, and he integrated Kombucha with this programme. His healing methods were so successful that they were adopted by many doctors. A company which carries his name today manufactures the Kombucha beverage and Kombucha drops. When it was reported that Dr Veronika Carstens (wife of a former German President) was using Kombucha with all her cancer patients, the fungus beverage became a popular healing remedy. Dr Sklenar and Dr Carstens recommended the use of Kombucha to their cancer patients as a complement to other therapies.

When it became known that the U.S. President, Ronald Reagan, was suffering from cancer, reports regarding his treatment were lost in the larger issues of politics. According to researcher and author, Günther Frank, Ronald Reagan heard of Kombucha through the autobiography of the Nobel Prize author Aleksandr Solzhenitsyn. Solzhenitsyn was diagnosed as having cancer in 1952 and fully recovered at the hospital in Tashkent ('The Cancer Ward') in 1953. Ronald Reagan received a culture from Japan and reportedly drank a litre of Kombucha daily. His cancer was prevented from spreading further and Reagan was able to finish his term in office. Dr Robert E.Willner (USA) cites Kombucha as being helpful in his highly recommended book *The Cancer Solution*.

KOMBUCHA IN AUSTRALIA

Immigrants from Asia and Europe introduced the Kombucha fungus to Australia. Today, a number of health practitioners who have treated their patients for all types of illness with the Kombucha beverage have found that, not only did their patients feel much better after drinking it, but that when used in conjunction with other medicine, the healing results were greater.

A Kombucha pioneer in Australia, Mr Jose Perko from The Way of Life Sanctuary in Queensland, believes that Kombucha made with green tea has better therapeutic value than that made with the black tea which is generally used in other countries. By experimenting with different fruit and herbal mixtures, (one of which was a pawpaw leaf concentrate), Mr Perko has managed to keep Kombucha mixtures for long periods without the use of artificial preservatives or causing unnecessary souring of the Kombucha. In further chapters I will discuss in detail some of his methods. I would like to thank him for the information he has supplied for inclusion in this book.

THE MANY NAMES OF KOMBUCHA

There are many names for the Kombucha fungus and for the brewed drink made from it. An interesting fact is that many of these names incorporate the word fungus, even though it is not really one. I call it Kombucha fungus and regard it as just a name rather than a botanical description. Other commonly-used names range from algae fungus to magical fungus.

In various languages names such as 'Russian flower', 'Russian jelly', 'Russian fungus', 'Japanese fungus', 'Japanese sponge', 'Russian mother' and 'Indian wine fungus' point to its heritage. 'Wondrous fungus', 'Magical fungus', 'Heroic fungus', 'Fungus giving long life' and 'Gout jelly' are names also given to the fungus in various languages.

Kombucha has the effect of giving the drinker a general feeling of wellbeing and is usually considered to be a remedy for many illnesses. With the beverage itself, most of the names are linked to the actual taste, such as 'Tea beer', 'Tea wine' and 'Tea cider'. In France it has been given the name of 'Elixir de la longue vie' - elixir of long life - which once again links the name to the effects it has on the health of Kombucha drinkers.

A list of all the names I have found so far follows:
Algae fungus; Algentee; Brinum Ssene; Cajnogo griba; Cajnyj grib; Cajnyj kvas; Cembuya orientalis; Chamboucho; Champagne of life; Champignon Japonaise; Champignon Chinois; Champignon miracle; Champignon de la Charite'; Champignon de longue vie; China-Pilz; Chinesischer Teepilz; Ciuperca de ceai; Comboucha; Combucha; Combuchaschwamm; Combuchatee; Conbucha; Devine Che; Elixir de longue vie;

Fungojapon; Fungus japonicus; Fungus tea; Funko Cinese; Ganoderma japonicum Gichtqualle; Gift of life; Godly Tsche; Gout fungus; Grib; Haipao; Heldenpilz; Hero fungus; Hongo; Indian wine fungus; Indian mushroom; Indischer Weinpilz; Indischer Teepilz; Japa'n gomba; Japanese fungus; Japanese sponge; Japanischer Teepilz; Japanischer Combucha; Japanisches Mutterchen; Japanpilz; Japonskagliva; Jponskij grib;

K'un-Pu-ch'a; Kambuha; Kargasok Pilz; Kargasok Schwamm; Kargasoktee; Kocha Kinoko; Komboecha; Komboechadrank; Kombucha Elixir; Kombucha tea fungus; Kombucha; Kombuchagetrdnk; Kombuchamost; Kombuchawein; Kombucha-thee; Kombuchaschwamm; Konko; Kwas; Kwassan; Lingzhi; Magic mushroom; Manchurian Elixir;

Manchurian fungus; Manchurian tea; Mandschurisch-japanischer Pilz; Mandschurischer Schwamm; Medusentee; Medusomyces Gisevii Lindau; Miracle fungus; Mo-Gu; Olinka;

Red tea fungus; Reishi; Russian tea-vinegar; Russian flower; Russian fungus; Russische Qualle; Russische Blume; Russische Mutter; Russischer Pilz; Sakwaska; Symbiont Schizosaccharomyces Pombe-Bacterium xylinum; Tea fungus Kombucha; Tea fungus; Tea cider; Tea beer; Tea kwas; Tea mould; Tea plant; Tea wine; Teemost; Teekwa; Teewein; Teepilz; Teeschwamm; Teyi saki; Thee – Schimmel; Theebier; Theezwam; Tschambucco; Tsche of Kombu; Wolgameduse; Wolgapilz; Wolgaqualle; Wunderpilz; Yaponge; Zauberpilz; Zaubersaft; Zaubertrank

KOMBUCHA – FACT OR FICTION?

Reports dealing with healing successes are often generalised and passed on without the desired research. With medicinal herbs, for instance, the general perception is that they can only heal and do not have any negative side effects. In fact, what is a health remedy for some may prove to be quite the opposite for others. When I first received the Kombucha fungus from a friend, I was given an information leaflet about Kargasok tea. Unfortunately, I have not been able to trace the author, but I will pass the report on word for word:

'KARGASOK TEA'

"Approximately 60 years ago, a Japanese woman visited the region of Kargasok (in Russia) and was stunned to find so many healthy people who were over the age of 100. She actually met a man at the age of 130 who had married an old woman of over eighty who was still able to conceive children. The Japanese woman was fascinated by this and attempted to obtain the secret of the eighty-year-old woman who hardly had a wrinkle. She discovered that in every household, young and old alike consumed about approximately ¹/₂ pint (US) or one-third litre of Kombucha tea each day. To make this tea, the Japanese woman received a special yeast fungus and

instructions on how to use the fungus. She took the fungus with her to Japan where she started to duplicate it. Her friends were invited to drink the tea as well and she also passed the fungus on to them with instructions on its use. They passed it on to their friends. After having consumed the tea for a period of time, the people began reporting successful effects:

– *A man with a blood pressure of 210/120 was able to reduce it to 140/80.*
– *A young girl diagnosed with shingles was cured.*

   "In Kargasok, cancer and high blood pressure are unknown. In Japan, soon after this tea became the subject of TV and radio programmes, well over one million Japanese were consuming the tea. Eventually the tea found its way to Taiwan, then to Hong Kong and now travels around the world where it is passed on from one friend to another as a token of appreciation and love. This tea appears to be a miraculous remedy for many types of suffering. Researchers have discovered that the fungus has three basic elements, without which the body cannot function. Encouragingly, Dr Pan Pen from Japan reported as follows on the effects of the culture: "This tea –
– *clearly lengthens the lifespan*
– *is a health remedy against chickenpox and shingles*
– *reduces the formation of wrinkles*
– *discourages the formation of cancer*
– *prevents adverse menopausal symptoms*
– *restores visual acuity*
– *strengthens leg muscles*
– *heals arthritis*
– *enhances sexual drive*
– *heals sweaty feet, constipation, joint and back pains*
– *heals abscesses*
– *heals blocked arteries and diabetes*
– *strengthens kidneys*
– *heals cataracts and heart disease*
– *restores the appetite and heals sleeping disorders*

- *reduces the chance of gall stones and liver problems*
- *reduces obesity and stops diarrhoea*
- *heals haemorrhoids*
- *helps restore colour to grey hair and improves baldness"*

This list may appear naive or overstated, and there is no record of the type of teas that were used, nor whether the tea brew or the fungus itself was being used. One cannot generalise about these conditions: one individual may be helped and another not. There is evidence that Kombucha helps restore hair. As for preventing hair greying, for me it has not worked, but perhaps it will for others. If you read the list of testimonials at the back of this book, you will be astonished at the wide variety of conditions that have been helped.

Can Kombucha be considered a miracle healing medium? Is it a remedy against cancer? Research will undoubtedly determine this in the future.

WHAT IS KOMBUCHA?

Kombucha is composed of a number of bacteria and special yeast cultures in a symbiotic relationship. This living organism ferments sweetened tea to become the Kombucha beverage. The refreshing taste gives one a feeling of wellbeing. The amount of living yeast it contains gives the Kombucha beverage an active life which continues after decanting it into bottles. Kombucha has become increasingly highly regarded as an aid to preventing many illnesses.

Part II:
HOW TO MAKE YOUR OWN KOMBUCHA

TO BREW YOUR OWN KOMBUCHA YOU'LL NEED:

INGREDIENTS:

USA

10 cups (4-5 US pints) of water
¾ cup white granulated sugar*
2-4 teaspoons (or teabags) tea – black, green, or a mixture
1 healthy Kombucha culture
2 tablespoons vinegar (This is used only with the first brew, if no starter Kombucha tea is available. In subsequent brews, use 1 cup (½ pint approx.) of the brew as a starter and omit vinegar).

Metric/Imperial

2 litres (3½-4 pints) of water
160 grams (5½oz) of white granulated sugar*
2-4 teaspoons (or teabags) tea – black, green, or a mixture
1 healthy Kombucha culture
2 tablespoons vinegar (This is used only with the first brew, if no starter Kombucha tea is available. In subsequent brews, use 200ml (⅓ pint) of the brew as a starter and omit vinegar).

Alternatively, a herbal tea - though not one that contains oil - may be used.* We recommend you start brewing with black and/ or green tea. When you are more confident you can experiment with herbs.

---

*Reduction in the amount of sugar recommended or substitution of brown sugar could prejudice the health of the fungus.
*You need to be sure that the herbs have been properly dried and stored without risk of spore or fungal contamination; those that contain appreciable amounts of oil are: angelica, bergamot, camomile, caraway, cumin, dill, fennel seed, lavender, lovage, marjoram, mint, peppermint, rosemary, tarragon, thyme, wormwood.

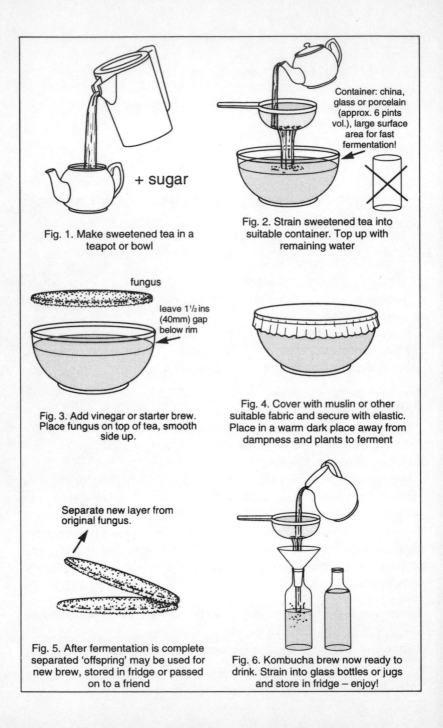

+ sugar

Fig. 1. Make sweetened tea in a teapot or bowl

Container: china, glass or porcelain (approx. 6 pints vol.), large surface area for fast fermentation!

Fig. 2. Strain sweetened tea into suitable container. Top up with remaining water

fungus

leave 1½ ins (40mm) gap below rim

Fig. 3. Add vinegar or starter brew. Place fungus on top of tea, smooth side up.

Fig. 4. Cover with muslin or other suitable fabric and secure with elastic. Place in a warm dark place away from dampness and plants to ferment

Separate new layer from original fungus.

Fig. 5. After fermentation is complete separated 'offspring' may be used for new brew, stored in fridge or passed on to a friend

Fig. 6. Kombucha brew now ready to drink. Strain into glass bottles or jugs and store in fridge – enjoy!

In this case, use 2-3 teaspoons of herbal tea, pour on boiling water, leave for 5 minutes, strain off tea leaves (or remove tea bags) and proceed with recipe below.

METHOD:

To make a batch of Kombucha, you will need a bowl made of china, porcelain, glass or ceramic that will hold about two and a half litres or six pints. Place the tea in a large pot, pour on boiling water, add sugar and stir until dissolved, and leave to brew for 10 to 15 minutes. Strain brewed tea into the bowl. Add the remainder of the water and allow to cool to room temperature, then add the cider vinegar (or 'starter' brew). Now place the fungus to float on top of the liquid. The fungus has a smooth side (possibly lighter in colour) and a rougher side. It should be allowed to float smooth-side upwards. A gap of at least one and a half inches should be left between the fungus and the top of the bowl.

Cover with muslin (or some other suitable cloth which will allow air through) and anchor this below the lip of the bowl with elastic.

The container should then be put in a warm place (ideal fermentation temperature is between 70° and 84°F (23° and 28°C), depending on the season). The fungus does not require any light but needs warmth and air. Smoke is harmful. After 6-9 days of fermentation (it is faster in summer or in higher temperatures), remove the fungus with clean hands, strain the beverage and pour into bottles leaving sufficient air space. (Alternatively, the beverage can be poured into one or more lidded jugs). After some experience you will decide the best fermentation time for *your* conditions. The drink should have a zingy and only slightly sweet taste, not too acid.

The bottles should be placed in the fridge, otherwise the fermentation process will continue and the beverage will obtain a sour taste. Once the fungus has been removed from the tea-brew, it can immediately be used to start a new batch.

The most suitable containers for brewing are bowls made of glass, porcelain or glazed pottery. Metal containers - including stainless steel - are not used because acids in the brew react with the metal. A comparision can be made with whisky or wine which taste different when fermented in wood. Plastic containers are more and more commonly used, in which case, they should be high quality food grade and acid-resistant. Polyvinyls, poly-propylenes and cheap plastics can cause chemical reactions in the brew due to leaching of plastics. Beer-brewing containers used for home brewing can be used. The containers should have a wide opening, not too tall nor filled up to the top. A wider, more shallow pot enables the Kombucha to ferment quicker and better.

THE FUNGUS REPRODUCES ITSELF

With each brew, a new fungus will have grown (by binary fission) on top of the original 'pancake' floating on top of the liquid. The pancakes can be gently eased apart by hand. The new fungus can then be used to start another batch or passed on to a friend. If no batches are started on the same day, the fungus can be preserved by placing it into an air-tight container with some Kombucha beverage, leaving an air space between the liquid and the lid and keeping it in the fridge until required. The recommended amount to drink daily is three average-sized wine glasses, one before breakfast and one 20 - 30 minutes after lunch and your evening meal. Larger amounts can be consumed quite safely for reason-ably healthy people. There are no limitations to the ways in which a Kombucha brewer can experiment with his beverage. The more experienced he gets, the better the resulting drink is likely to be – Cheers or Good Health!

CONTINUOUS FERMENTATION PROCESS

After I first wrote about Kombucha, I received a mass of com-ments and enquiries from people in Eastern and Middle Europe.

A traditional pot for continuous fermentation. Note the tap is above the yeast settling level.

These people recalled their mothers or grandmothers having a large container of between 10 - 20 litres, (3 - 4 US gallons) stored in a warm place.

People drank the brewed beverage when required and, every now and again, the container was refilled with 4 to 5 litres (10-12 US pints) of sweetened tea. This was one form of continuous fermentation. Different top-ups can be made with different teas. Children, for instance, often love a mix made with wild blackcurrant herbal tea. I have never heard of any problems with this type of fermentation. In counties where Kombucha drinking has been common for generations, children drink it without restriction.

On reading about this, I immediately purchased a large container, with a tap located several centimetres above the base, so that it would not discharge the yeast sediment from the bottom. I discovered that Russian and Polish workers had brought this process with them into the Eastern regions of Germany. Some war refugees also brought this process with them to the West.

Continuous fermentation, besides being easier, also has the advantage that a larger amount of fluid does not react so quickly to temperature variations. A large container can also be equipped with heating plates as used with the fermentation of beer. In my experience, I must say that the continuous fermentation process is the ideal way to brew Kombucha for personal use.

SOME FREQUENTLY-ASKED QUESTIONS ABOUT BREWING KOMBUCHA, Compiled by a healthy Kombucha drinker and producer in England:

How long should I have to wait to drink my Kombucha?
A week from brewing is an average time. Six to eight days after storing your covered bowl of Kombucha, taste the tea. It should be neither too sweet nor too sour. If it is too sweet, allow the fungus to ferment for another day or so. If the taste is rather sour, then leave the next brew for one day less. You may still drink the tea – it will be very good for you, though you may like to add some

spring water or fruit juice!

What is the correct temperature for the mushroom?

The Kombucha mushroom likes the sort of temperature used in wine or brew fermentation. For that reason, if your room is cold (in winter, for example), it might be useful to use a heated brewing mat or thermal collar. These can be purchased from most brewing shops. In summer, average daily room temperatures will be fine. An airing cupboard is a popular place in which to place the brew. However, if the hot water tank is not well-insulated, the temperature will be a little high. Between 73°-83°F (23°-28°C) is about right.

How do I know if it is working?

Firstly, don't be surprised if the fungus sinks to the bottom. It generates carbon dioxide under its surface as it transforms the tea and may need a while to build up sufficient gas and buoyancy to float back to the surface. After a few days, you should notice a smell of vinegar and the formation of a transparent membrane on top of the fungus. If this occurs, all is well, this is the new fungus forming.

What happens if I damage the mushroom?

The fungus is very hardy and will still produce a well-formed offspring, even if torn or partly-formed. The offspring may be thinner over the damaged area, but the thinner area will get thicker with successive generations. If the accumulated gas under the mushroom pushes the thin part of the fungus membrane away from the tea surface, you may lift it to allow the gas to escape. This will bring the thinner membrane back into contact with the liquid surface and will encourage a more uniform growth.

What happens if I go on vacation for a while?

The Kombucha fungus is a hardy beast and will sleep quite happily in a little of its tea fluid for three to six months when left in the refrigerator. It's best to lay it out flat and make sure that it has a little air in the container so that it can breathe.

A typical two litre bowl with muslin cover secured by elastic.

How do I produce more cultures for my friends?
A Kombucha will happily produce offspring in just a small (half-an-inch or so) amount of tea. So, in this way, you can produce more offspring without using lots of sugar and tea and without having to generate large amounts of Kombucha tea. The resulting Kombucha tea may be a little too sour to drink, however, so prepare at least one batch with the usual amount of liquid.

Once you have a number of fungus pancakes, it is possible to refrigerate them successfully, stacked in a container with a little fluid in the bottom. Kombucha offspring do not fight - they live very happily together! All the mushrooms may be kept active by brewing from them in rotation. In this manner, you will always have a number of healthy, active mushrooms to give away on request.

What effect will drinking the tea have on my health?
The effects of drinking the tea vary from person to person. In general, however, the tea acts in a gentle, balancing manner and helps the body heal itself in whatever way is necessary.

As a general rule, the more in touch you are with your body, the greater the effect you will notice. You may feel warmth, notice an increased brightness of vision, a general relaxing, an opening of your energy. A general feeling of wellbeing is often reported. The number of colds you get may well be reduced. Older people frequently report an increase in mobility and a decrease in niggling joint pains.

If your body is already very open and relaxed and your energy strong, you may not notice these effects. You may simply register Kombucha as the tasty, refreshing, slightly fizzy, benign drink it is, which helps you to stay well!

PROBLEMS ASSOCIATED WITH KOMBUCHA FERMENTATION

With any living fungus, accidents can occur which may prove harmful or even lethal to the organism. If the temperatures are not regulated or if the fungus has been stored for a long period of time, it can acquire a glutinous coat. If this happens, it should be cleaned, using lemon juice or wine vinegar. It can then be used again. It is a little harder if the fungus has started to grow mould. Too low temperatures or an unclean environment can be the cause of this. In this case, dispose of the beverage and wash the fungus as described above. The usual 10% should be taken from another brew to start a new batch. If there are no other batches, put the fungus into the prepared tea with the required sugar content and add two tablespoons of vinegar (better still, Kombucha vinegar if available). The beverage should have a sour smell to it after a few days. This is an indication that the fungus is working. It shouldn't have a mouldy smell. In areas where the vinegar fly *(Drosophilia fenestarum)* is active, it is very important to cover the beverage with a muslin cloth fastened by a rubber band to prevent the fly entering the container and laying eggs. *(Meixner).*

Don't wash the fungus if there is no obvious reason to do so, unless long storage or mould indicates it to be necessary. The fungus is always slimy! In one case I came across a fungus which did not work correctly. The person who owned it had apparently done everything right. I went to the supplier in the next town and found that the liquid used had been stored in a 9 US gallons (25 litres) container with a narrow neck. It had been filled right to the top with the fabric cover touching the fungus. The fungus simply did not have enough air to work properly. Even by doubling the fermentation time, the brew was still too sweet and the drink not healthy. I traced the culture further back to the previous supplier where it was not doing well either. Brewing had been done here for the previous five months with only 20 grams of sugar (less than an ounce) being used per litre (1¾pints). All these cultures were brought back to normal performance. With two of them, I added commercial beverage and with another I used a double strength black tea with double sugar and allowed it to ferment until very sour. With this brew, normal fermentation with good results continued. See also p.41. **NB: A dead or ailing culture cannot perform properly and will not produce a healthy drink.**

NICOTINE: A DEADLY POISON FOR THE FUNGUS

Tobacco smoke will kill the fungus. People have often rung me and told me that their fungus was not working any more. In 50% of these cases I found that someone was smoking in the room where the fungus was fermenting. The fungus cannot tolerate tobacco smoke. With occasional chimney smoke the fungus may survive. The beverage will, however, adopt a smoky taste which will still be noticeable in future batches. Tobacco is an example of how man can develop unhealthy products from nature. However, the negative effects of tobacco can be used to advantage by brewing tobacco tea and spraying in your garden as an effective insect repellent! Under no circumstances should the tobacco be left in a food container, and brewed as tea by mistake, which could cause death if drunk. This happened when I was a child and one of our neighbours grew tobacco in his garden, storing the dried leaves in his kitchen. The family had a large variety of herbal teas and, by mistake, the tobacco was used to brew tea, with deadly consequences.

A SEPARATE FUNGUS IN CASE OF EMERGENCIES

As with all living organisms, the Kombucha fungus can develop problems which render it useless or dead. It is, therefore, wise to have a separate fungus stored in the refrigerator to ensure the continuation of Kombucha brewing in case of an accident. The fungus stored should be from a healthy specimen, with which good brewing results were achieved. The fungus should be placed in an airtight container together with approx. ¼ litre (½ US pint) of brewed Kombucha beverage. The fermentation liquid will sour very slowly in the refrigerator and will be as sour as vinegar after three to six months. When starting a new batch with this fungus and vinegar, you will find that the fermentation process will be accelerated and the fungus will develop a new layer much quicker.

WHAT TO DO IF A FUNGUS SHOULD GET TIRED OF REPRODUCING

It can happen to every Kombucha brewer – the fungus will grow tired due to something in its environment and will lose its full strength and ferment the tea slowly. While the fungus may still work and ferment the sugar, the beverage may taste flat. The brewing process should then be continued using green tea. Additionally, it helps if a beverage is used from another, very lively fungus to restore the tired fungus. *(Perko).*

KOMBUCHA, AN ALCOHOLIC BEVERAGE?

When brewing 2 litres (4-5 US pints) of Kombucha, sugar is fermented, which produces alcohol. The amount of alcohol depends upon the temperature and amount of sugar used. With a minimum sugar content of 50 grams (2 ounces, ¼ US cup) the beverage will have 0.1% alcohol after 14 days and approx. 0.3% alcohol after 21 days. By using larger amounts of sugar (200 grams, 7½ ounces, 1 US cup), the beverage can have up to 2% alcohol after 14 days. At the same time, however, the beverage will have a very sour taste and be practically undrinkable. With an ideal amount of sugar (160 grams, 6 ounces, ¾ US cup) and a fermentation time of between 6 and 10 days, it can be assumed that the alcohol will be of very small proportions. Pastor Weidinger states that Kombucha has very small amounts of alcohol, no more than 0.5%. The beverage can, therefore, also be given in sensible amounts to children and teenagers. Muslims and Buddhists drink it without concern. "Recovered alcoholics do not have to fear the small amounts of alcohol." The Salvation Army is using Kombucha to help alcoholics.

Interesting in this context is a letter from Mrs M.T.: "I tried Kombucha with blackcurrant and it tasted great. I no longer need that alcoholic drink to relax when I get home after work. Kombucha is better - it doesn't cause aches and pains, and it doesn't keep me awake at night."

CAN THE FUNGUS BE EATEN?

This question emerges time and time again. I ask myself, 'Why would anyone want to eat this fungus? It is as tough as leather and slimy!' I do not find it very appetising. However, the fungus is not harmful. Only once in any literature dealing with Kombucha did I find a specific reference to eating Kombucha, from Professor Lindner, a German researcher, who observed that the slippery mass easily passes along the bowel walls and aids in constipation, but see p.95. The fungus itself has all the healing properties the beverage has, but in a very concentrated form. A pressed extract can be obtained from the fungus and is sometimes sold in health food stores. If the pressed extract is a concentrated medicine, why indeed should one not eat the fungus, if one finds it appetising?

LIFESPAN OF THE FUNGUS

If properly treated the fungus will last for some months. On the surface, it constantly builds new layers. In turn, on the underside, the layers die off. If the fungus is well cared for, it is a gift which can be given from generation to generation.

KOMBUCHA - PRESSED EXTRACT FOR DIABETICS AND WHILE TRAVELLING

Late in the 1920s, the Kombucha researchers Wischofski and Hermann produced it in a concentrated form. The idea was to achieve a product in pure form which would be available with a certain consistency. This was important for being able to make comparisons while doing research. Known as Kombucha drops or 'Kombuchal', this product was patented in Germany and was available in pharmacies. The drops were manufactured by vacuum distillation of the liquid culture. Apart from alcohol and acetic acid, Kombuchal has all the components of normal Kombucha. From the records of the internal clinic in Prague it was determined that all tests showed results favourable towards old age and arterial blockage.

The idea of pressing the Kombucha fungus came from its most famous researcher, Dr Sklenar himself. The resulting products were named 'Drops Kombucha D1' (D1 being the homoeopathic dilution degree of 1 to 10) and Kombucha tincture. With the drops he achieved healing results which gave him a reputation that lasted long after he died. People who do not wish to miss the energy-restoring and digestive effect of Kombucha are advised to take Kombucha extract with them in this handy form when travelling. It can help with travel sickness.

Kombucha is recommended for diabetics, although the sugar content cannot be determined when 'home brewing'. Although experts say that Kombucha does not have any negative side effects for diabetics, I suggest that it would be wise to use some of the more sour-tasting products. Kombucha drops, or pressed extract, are highly concentrated forms of the effective parts normally found in the beverage. Manufacturers recommend the use of 15 to 20 drops, three times daily for adults, to be taken with water.

STORING KOMBUCHA AFTER FERMENTATION

The Kombucha beverage is a living thing, hence its healthy and energising effect. Kombucha fermentation occurs more quickly at room temperature and slows down considerably in the refrigerator. If left to ferment too long, the end result is a sour beverage that is still healthy, but not to everyone's taste. During slimming diets, this sour beverage is used frequently. Caution should be taken with bottles stored for a long period, however; they may explode due to a build-up of gas.

Commercial Kombucha beverage, available in some health shops, may have a very sour taste, which would be a good indication that it had not been preserved or tampered with. The nutritious effect of this beverage is just as good as 'home brew'. It is possible that, when first tasting commercial or over-brewed Kombucha, it may be unappealing and sour to some palates. Kombucha should taste pleasant (apart from some herbal tea combinations) and one should not forget that unlimited different flavours can be achieved by experimenting. Depending upon the length of the fermentation, the liquid will be either clear or misty. After a short period of time, the yeast will settle on the bottom. This yeast sediment can either be consumed, as with 'wheat beers', or filtered off with a cloth. 10% of the liquid should be used to start the next batch. The rest may be filled into bottles and sealed with corks - taking care to leave a little gap at the top to prevent the bottles from exploding.

DOSAGE, HOW MUCH CAN ONE DRINK?

Pastor Weidinger has had experience with Kombucha for well over fifty years including brews from such countries of origin as Formosa and Taiwan. To this question, he has made the following comments: *"All natural healing remedies give life. Living organisms are being given into our living body, absorbed, processed and sorted through our blood circulation. When I drink something it is not just merely 'open wide and down the hatch'! No! Much more happens. For one thing, the beverage becomes part of myself. I react to it usually in a way quite opposite to what I expect the reaction to be. First of all Kombucha will effect our soul, our feeling. We will feel happy, relaxed, satisfied. Then our consciousness will react; the ability to absorb. We find it easier to concentrate, to make decisions. We are less likely to forget important issues. It affects the activities of our mind, our intelligence. Last but not least, our body will react. So how do we learn to drink Kombucha properly? Each individual must discover his or her own answer, as every individual acts differently, feels differently. You are an individual, with you everything functions your own way. The groundwork has been laid by our Creator. You do everything differently from other people. A lot may be similar, but not everything is the same. This fact is very important to consider when deciding how much Kombucha one should drink.*

"After consulting with a doctor, you could drink an eighth of a litre daily after the main meal to help the digestion. This should be based on an exact time plan with rest periods."

One circular letter I found being distributed with Kombucha mentioned some possible negative side effects:

"Rapid detoxification can cause discomfort in some people if too much tea is drunk when beginning treatment. Work slowly up week by week from six tablespoons daily in divided doses. At two table-spoons, taken three times daily, discomfort can usually be avoided. Some possible side-effects can include headaches, stomach aches, nausea, fatigue, dizziness, mild diarrhoea, constipation, pimples, rashes and wind. However, these are temporary effects lasting from a day to a week or so, in basically healthy people. Drink extra water to counter them. People with a serious disease condition may experience a healing crisis if they drink too much too soon and may like to begin with even less – say 1-2 teaspoons three times a day, gradually increasing the amount. Kombucha effects appear to be-gin in the weakest part of the body, then the second weakest and so on."

THE PROPERTIES OF THE KOMBUCHA BREW MAY CHANGE WITH THE FOLLOWING:

1. The culture can change and cease to perform correctly. If, after a certain period of time, the brew still tastes sweet, the culture is not functioning. This can happen when the culture has been overheated in transport, for example, or on the cooker (one person I know put the brew on the stove for temperature control-led brewing!) See also p.25.

2. The type of tea used (green tea, herbal tea, etc.). This part of the Kombucha brewing process results in most of the likely prob-lems. Negative side effects are in my opinion not due to the culture itself, but can be traced back to the **additional ingredients** and the fermentation process. The herbs, green tea and black tea

included, are transformed in the brewing process. Their properties are amplified during fermentation and more easily assimilated and utilised by the body. (See also pp.55-62.)

3. The amount of sugar used, the time of fermentation, the average fermentation temperature and the time and temperature after the main fermentation in the fridge; any or all of these factors may have an effect.

MIXING TEAS

The most popular Kombucha drinks are the ones in which different tea mixtures are used, according to taste, in the brewing during fermentation (see pp.57-62). Some people add fruits to the brew two days before the end of fermentation. This not only improves the flavour, it also adds vitamins and all the other benefits of the particular fruit or vegetable in the drink. When a seasonal fruit is used, it should be thoroughly cleaned and blended in the kitchen mixer. A strong fermentation is triggered shortly after the fruits have been added. The result is delicious. Have you ever tried raspberry-, mango-, strawberry-Kombucha? Your guests will like it, young and old!

In response to countless enquiries from people, I have to say that I have come across only a few negative effects. These are:

Constipation (one case). Black tea was used and the fermentation time was not long enough (the drink was still very sweet). Since black tea can cause constipation in some people, a mixture of buckthorn, elder flowers and knotgrass and raspberry was used instead, with good results.

High blood pressure (one case). Green tea had been used. A mixture using hawthorn and other teas was substituted with success. (See also pp.78-80.)

Candida albicans (a few cases). In these cases, too much brew had been taken too fast. (See also p.86.)

With negative side effects, one has to consider that there is no such thing as only one side. While most people have a very good result with Kombucha, there is always the slight chance that a few will have a negative reaction, whatever the reason might be. To test the possible negative side effects of Kombucha, I drank between two and three litres every day for six weeks. I felt great, but that does not mean that everyone can do the same. Trial and error is the only real answer.

EFFECTS OVER LONG PERIODS OF TIME

I believe that nothing should be used over long periods of time, as our bodies will get used to it and the required effect will not last. Tests have revealed that vitamin treatment is much more useful if adopted over limited periods of time, rather than continuously. It is also much more economical. In most homes where Kombucha has been consumed over decades, people usually take an occasional break of between one and two weeks. When under stress, or if there is a virus going around, they return to drinking Kombucha again.

How long should one drink Kombucha? Pastor Weidinger recalls that he once received the following letter: 'Through a friend I received a Kombucha culture and I have already started brewing. Is this drink for permanent consumption or should I stop in between? How can I keep the culture alive while not using it?' His reply was, *"The length of any therapy depends upon the motive. If I merely want to live a healthier life, I would recommend a three weeks programme, drinking a wineglass daily before breakfast, preferably starting after a full moon and ending one week prior to another full moon. If I wish to treat specific illnesses such as bowel problems, strengthening of the liver, rheumatism, gout or improvement of blood quality, I would recommend drinking three-eighths of a litre (3 wineglases) daily of Kombucha, over a period of between 3-6 months, then stopping for one month before continuing again. Permanent use is not recommended, as the body requires a break between treatments in order to make proper use of the help being offered. If one has a break between treatments, the user will feel the effect much better when starting again."*

This is called the 'Interruption Theory' which is now recommended by a number of Kombucha experts, to allow the body to consolidate a new balance.

THE TRANSFORMATION OF SUGAR IN KOMBUCHA

When they read, in the recipe, that one should use ¾ US cup (160 grams) of sugar per 4-5 US pints (2 litres) of tea, a lot of health-conscious people stare in disbelief. So much unhealthy sugar? The answer is easy. YES! The Kombucha culture requires sugar to ferment and stay alive. Through the fermentation process, sugar gets transformed into lactic acid and alcohol. With less than ¼ US cup (50 grams) per litre, the fungus will starve.

Pastor Weidinger answers the question in the following way : *"Sugar on its own may cause blood disorders and depression. As a sweetener, it has been rejected by our society for years now. Many experts agree that the constant use of sugar has a negative effect on the blood quality. This may then result in poor functioning of the liver and, in turn, influence our moods. So how about the use of sugar in Kombucha? As the sugar in the Kombucha beverage is completely converted, there will be no negative effects of any such kind."*

Recently a person (can of Coke in hand) tried to make me aware of the damaging effects of the sugar in Kombucha. She had problems in believing that in every Coke she drank there were not only seven teaspoons of sugar (unfermented!) but also 32mg of caffeine!

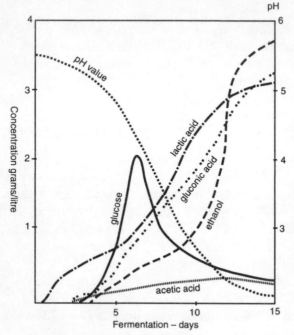

This graph shows how the sugar is converted in a brew made from black tea – initially into glucose which then reacts with the tea to produce the health-giving organic acids – making a drink which should not be particularly sweet. The duration of this process depends on the ambient temperature. (This research was conducted by Dr Jürgen Reiss and published in the German food review "Lebensmittelrunschau" in 1987.)

HONEY INSTEAD OF SUGAR?

The use of sugar continues to pose a matter of concern for many health-conscious people. However, the fact is that the fungus requires sugar to survive. It ferments into 'healthy' components. With artificial sweeteners, the fungus would starve and die. The effect of honey on the fungus would also be detrimental and cause it, eventually, to die. Sugar is essential for the fungus's own digestive process. This is also the reason why the amount of sugar does not necessarily make a difference to the taste, once the fermentation process is complete. Over one cup (190 grams) per US quart means that the amount of sugar remaining after fermentation will be too high, which may pose a problem for diabetics as well as people with cancer or stomach and bowel illnesses.

When conducting trials using honey myself some years ago, the fungus became tired and slowly died. I was surprised, however, when I met Mr Perko from The Way to Live Sanctuary in Queensland, to learn that he had successfully brewed Kombucha using honey instead of sugar. Mr Perko has attempted more trials brewing Kombucha with herbs than any other person I have met. He is making further experiments to assess the effects of various types of pollen in the honey which may have an effect on the Kombucha culture. With the mixtures of various teas that Mr Perko has tested he always used elderflower tea. Elderflower tea can be used to start a Kombucha fungus. Elderflower has the effect of giving the Kombucha beverage a certain champagne character, a typical example of the many different achievable varieties of Kombucha. Mr Perko has an extraordinary feeling for and skill with the brewing of Kombucha. For normal use, however, I would strongly advise against the use of honey when brewing Kombucha.

If you do decide to try using honey for Kombucha brewing, a temperature of 83°F (28°C) is very important. This is why experiments in tropical Queensland have been more successful. Mrs J.G. from Queensland has devised a simple method which has worked very well over a long period of time. She uses honey in every second brew, so that the individual fungus ferments one week with honey and the next week with sugar. With several containers brewing simultaneously, she always has a honey brew which is mixed with the conventional brews and produces an excellent tasting drink.

Mr Bill Corrigan from Queensland has also written of his observations of and experiments with Kombucha brewed with honey: *"My own rational mind could not accept the judgment on honey, particularly when you consider that cane sugar has only become prominent in recent history and would not have been used or known in most cultures that used Kombucha. The only problems I had were in the colder months when temperatures varied. When there are uniformly warm temperatures – no problems. I always brew using green tea and some herbs separately, and allow the mixture to cool until warm before I add the honey. The reason for this is that no enzymes are lost to the hot liquid. The amount of honey – no less than four tablespoons of honey per litre, preferably more."*

Bill uses fresh herbs from his own garden like nettle, dandelion, yarrow and elderflower, which he adds to the brew two or three days before the end of fermentation. He stores Kombucha in bottles for two and a half months at normal temperatures under his house in tropical Queensland. He was delighted to find the taste improves with age, without bitterness.

For brewing with honey, temperature-controlled containers are necessary in cooler areas. That is the reason why my own trials did not have the desired results and why publications from cooler countries say a strict NO to honey brewing.

KOMBUCHA VINEGAR

It has probably happened to a lot of Kombucha brewers that one batch has been forgotten and has continued to brew over many weeks. The result is a Kombucha vinegar which can be used like any other vinegar, but can also be used to clean the fungus should it have become mouldy.

PRESERVATION OF KOMBUCHA

If you treat your fungus properly you will have the joy and gift of receiving health and enjoyment for life. The fungus itself is a very stable unit. With temperatures above 65°F (18°C) it will ferment, provided it gets enough sugar and oxygen. If you have too many individual fungi or go off on vacation, the fungus may be preserved by placing it in approx. ½ US pint (¼ litre) of its own liquid and storing it in the refrigerator. The fungus then falls into a dormant type of winter sleep. The container in which the fungus is stored should only be half-filled, so that the fungus will still be able to breathe. Storing it for a period of three months in the refrigerator is possible without any problems. Whether the fungus can be frozen is debatable. Some experts say there are no problems with freezing the fungus but others, such as Dr Meixner, contend that the frost crystals could be harmful to the fungus structure. Trials may have been conducted with different cultures. Personally, I do not see any reason why it should ever need to be frozen since the fungus can survive for a considerable period of time in a refrigerator.

Drying the Kombucha fungus is another way of conserving the culture, especially before shipping. The dried fungus is very light and can be sent quite cheaply by airmail. The usual containers used for storing the fungus are likely to be damaged in transit and leak. The most favourable drying temperature is 90°F (33°C). Drying in direct sunlight or in the microwave oven will kill the fungus. Dehydrators, available from hardware store and commonly used for fruit and other foods, are ideal. When starting a brew with a dried fungus, use half a litre of tea at double strength for the first brew rather than the normal concentration for Kombucha brewing.

Here's a suggested recipe:

Ingredients:
1 litre (5 US cups) water
160 grams (5½ ounces, ¾ US cup) sugar
2 teaspoons (or teabags) black or green tea
one dried fungus
if available, ½ US pint (200 mls) commercial Kombucha beverage, or one tablespoon (15ml) boiled vinegar

Method:
Pour the sugar into the heated water and stir until dissolved. When the water boils, remove from the heat and add to the tea. Leave to brew for ten to fifteen minutes.
Cool the tea until lukewarm and strain into a glass, porcelain or pottery container. Add the dried fungus.
Add commercial Kombucha beverage (if available), or one tablespoon of vinegar. Cover and store in the usual way, making sure that a gap of an inch or two is left between the top of the container and the cover.

A temperature between 73° - 83°F (23° - 28°C) should be

maintained for the first week. Ferment this first brew for at least fifteen days until it tastes very sour. With this brew, a bigger batch of two or three litres can be started. You will find details about further brewing method in pp.19-22.

KEFIR, YOGHURT AND SIMILAR CULTURES

I am continually being asked whether Kombucha can be compared to Kefir or ginger culture. With Kefir, Kummis or yoghurt, cultures are used for the fermentation process of milk. The most popular of these three is yoghurt, which has been available for well over fifty years. Natural yoghurt (without the addition of any fruit or flavours) can easily be produced at home. The micro-organisms in yoghurt (*Lactobacillus bulgaricus* and *streptococcus thermophilus*) ferment milk within three hours when kept at 110°F (44°C). Fruit can then be mixed in if desired.

I am asked again and again about ginger culture used for the manufacture of ginger beer. My research shows that ginger beer was brewed in wooden barrels years ago and a white froth on the top added a good taste. With the brewing methods of today, ginger beer is fermented with the beer yeast called *Saccharomyces cerevisiae*.

Tibi (sometimes called Kefir) is another culture similar to Kombucha, using water in the process. Tibi appears originally to come from the Caucasus. The culture consists of small white transparent pellets which, when cultivated in water, increase at a rapid rate. The fermentation time is approximately two to three days. The gas content increases day by day. Similar to Kombucha, Tibi requires sugar to live. Sultanas and dates are added to the fermentation process as an aid and to improve the taste. Tibi is also known as Japanese crystals and Water Kefir. Tibi beverage is generally used to improve water, and in this way, overall health. It aids the body in the detoxification process, which in turn helps stability. The recommended dose per day is approximately one litre of Tibi.

KOMBU-TEA

The use of the name 'Kombu' in Japan is rather confusing. Reports about the health benefits of 'Kombu-cha' ('cha' stands for tea!) from Japan are probably based on the sea-vegetable-tea Kombu (kelp = *Laminaria* family). 'Kombu-tea' is harvested in salt water and has nothing to do with the tea-fermenting fungus 'Kombucha' which grows in sugar water with tea.

I have discussed Kombucha with Japanese people who have used this term interchangeably for the salt water 'Kombu-tea', before I realised we were talking at cross purposes.

Whether Kombucha is spelt with K or C does not really matter. Due to the intensive research conducted by famous doctors such as Dr Sklenar, Dr Wiesner and Dr Carstens, it appears that the name Kombucha dominates. The German bioculture expert, Dr Alex Meixner, calls it Combucha in his book *Pilze selber züchten (How to grow mushrooms)*; at the same time he issues a warning that all Kombucha products are not of the same quality.

A healthy culture.

Part III:
HOW DOES THE FUNGUS WORK?

KOMBUCHA FUNGUS – A BIOLOGICAL PRODUCTION CENTRE

Today we consume vast quantities of preservatives, cook with microwave ovens and use other processes which destroy the living energy in our food. It is important for our digestion to consume living micro-organisms and bacteria (though the thought may not be very appealing!). Without these tiny micro-biological helpers, however, we and plants would not be able to survive. For example, approximately 250 million bacteria work in one gram of good gardening soil. Wine and beer would not exist without micro-organisms. It is a miracle that these invisible helpers can adapt so easily. They seem to have been able to survive quite well in our chemical age with its approximately 3000 food additives, though we should not be surprised that new illnesses and allergies are created with this bombardment of foreign products and additives.

Pastor Weidinger comments as follows: *"The breeding does not happen through spores as with conventional yeast cultures. Kombucha is a lichen. Lichens are regarded as the oldest food and healing components of man. These remarkable plant organisms evolved some 2·5 billion years ago. It began with sea algae which were confined to the space of the ocean. When they moved onto land, they were dependent upon a mate, which they found in the fungi ('mycel'). They formed a living relationship and evolved into one unit, the lichen or lichens. Only in this form were they able to survive. The algae supplies the fungus with organic nutrients, carbohydrates. The fungus had the function of serving as a water reservoir also containing vital minerals."*

"There are well over 1600 types of lichens. The manna lichen delivers food for man, the reindeer lichen supplies food for

animals. *Others such as Irish moss and lung lichen act as healing and medical components, the Oak moss for the manufacture of perfumes. Lichens incorporate strength. They seem to be messengers from God which came to us from another world. Although there is a large variety of them, they all have the same basic principles, the lichen acids. These are vital components of the lichen which revitalise living matter, minerals and trace elements. Where there are lichen, life is awakened. The tea fungus Kombucha is more than just a fungus; it is a symbiosis of algae and fungi. Kombucha is a lichen. Kombucha tea is life, receives life, aids life. A fully grown Kombucha lichen builds a gelatin-like tough membrane, assuming the same diameter as the container in which it grows.*"

The fungus expert, Dr Meixner, writes that the important components of the Kombucha fungus are the tropical yeast *Schizosaccharomyces pombe, Saccaromycodes ludwigii* and *Pichia fermentans*. The most important bacteria in the symbiosis is apparently the slimy vinegar bacteria *Acetobacter xylinum*, which creates a slippery mass out of cellulose in the symbiosis. These bacterial and fungi parts are bound in the tea fungus, therefore enabling symbiosis.

Dr Helmut Golz mentions the following five active **bacteria** in his book '*Kombucha - an old tea health remedy*'
* *Acetobacter xylinum*
* *Acetobacter xylinoides*
* *Gluconobacter bluconicum*
* *Acetobacter aceti*
* *Acetobacter pasteurianum*
    and as **yeasts** he mentions:
* *Schizosaccharomyces pombe*
* Yeasts originating from *Apiculatus*
* *Saccharomycodes ludwigii*
* *Pichia fermentans*
* *Mycoderma*
* Yeasts originating from *Torula*
    The split yeasts *(Schizosaccharomyces)* increase in contrast

with conventional yeasts by cross-division. This appears to be the reason why Candida albicans sufferers can enjoy yeasty Kombucha beverage *(see Frank)*.

KOMBUCHA CULTURES ARE NOT IDENTICAL. IS THERE AN AUTHENTIC KOMBUCHA?

There is no exact answer to this question. Kombucha cultures have existed on this planet for thousands of years. These tea fungi were grown in conventional kitchens and have different stocks or combinations of yeast and bacteria. I think the easiest comparison can be made with apple trees. In an orchard, there are twenty apple trees which look the same. Which is the best tree? This question can only be answered at the time of harvest. Some trees will have sour apples, some sweet ones, some juicy. The continuation of the stock will only be done using the best. Pastor Weidinger explains his selection process as follows:

"Only the best mother animals and mother trees will produce good stock - a fact of which every grower is aware. The important factor is to reproduce something living that has its own unique qualities and cannot be exchanged with something else. The same goes for reproducing a Kombucha fungus. Only a large solid and well-developed Kombucha fungus will be capable of being divided."

Research into the origin of the culture shows that it has been grown for over two thousand years in stone jars in unsterile farm kitchens. Thanks to its antibiotic resistance, it has retained its health and stability. Is the laboratory culture, grown hygienically for 30 - 40 years, still as stable? During brewing, the culture has to be kept clean, otherwise mould will build and the culture will die. In some cases the culture will still ferment, but the beverage will not be so lively and the effect will not be the same as that of a healthy culture. It is important to continue brewing with a fungus that has the best qualities. In case an 'accident' happens and the fungus does not work as it used to, it is helpful to put the fungus into 'mother tea' to help produce the missing bacteria and yeast components.

The effect of the varying beverages can be compared only when their conditions of culture are the same, that is when they include:

* the same type of culture
* the same sort of tea (green tea, herbal tea, etc.)
* the same amount of sugar per litre
* the same time of fermentation
* the same average fermentation temperature
* the same time and temperature after the main fermentation in the fridge - during which the beverage will still continue to ferment.

Variations of the above can achieve many different taste directions and effects on performance. Different combinations may cause problems for some people. It is therefore particularly important to use only the best and healthiest cultures for further brewing and multiplication. A fungus which is not working properly should be discarded. Kombucha is the tea with one thousand and one different tastes. The purchase or gift of a Kombucha culture requires trust. If you need help in obtaining a proper culture, please contact the address mentioned in the appendix.

CAN KOMBUCHA BE CONSIDERED A MIRACLE HEALING MEDIUM?

When I hear terms such as miracle healing medium, it normally makes me rather sceptical. Just as there is never only one cause for any particular illness, so it is unlikely that one general miracle medium exists. There is agreement that Kombucha balances the digestion and helps regularity. 'We are what we eat' is an old saying. Hippocrates said: 'Death sits in our bowels', and links many illnesses to problems with the digestion. According to experts, the healing of the bowel flora is the first and most important step to curing an illness. In traditional medicine, the application of

Kombucha showed such positive results that words such as magic and miracle were associated with it.

THE LACTIC ACID FERMENTATION

The term lactic acid fermentation does not sound very appetizing. However, the products associated with this term are not only well known but also very popular. What would we do without wine, beer and cheese, on our menu? The lactic acid fermentation process is an important part in preservation. One of the best examples of this process is sauerkraut. The famous sea captain, Captain Cook, always took sauerkraut with him on his expeditions, as this was the only way to preserve vitamins in those times. Without the important vitamin C (approx. 20 mg per 100 grams of sauerkraut), the men on these ships would have contracted scurvy. Of course sauerkraut is not necessarily consistent over the years. Today salt is mostly used in sauerkraut, but in the early days people used herbs and spices such as juniper berry, mustard seed and thyme, instead of salt. To avoid calcium deficiency, it was common to add eggshell flour to the sauerkraut.

As a strengthening beverage for manual workers, a vinegar beverage was mentioned three thousand years ago. 'Come hear, eat your bread and dunk it into your vinegar beverage' (*The Bible*, Ruth 2,14). The farmer, Boas, used these words to invite Ruth, later to be his wife, for a meal. The advantages of lactic acid fermentation were already well-known in ancient times.

Sour dough bread is the basic food source in many countries. It keeps fresh longer and is easier to digest. Particularly interesting was a conversation I had with a Maori lady. I explained to her how to brew Kombucha, how to multiply the fungus and so on. She asked if this was similar to dealing with the bread-plant *(R'ewa'n'a P'arawa)*. I didn't know the answer. It was news to me that, at this end of the world, fermented bread was traditionally produced.

R'ewa'n'a P'arawa was handed down from generation to

generation as a religious gift. The bread was made from the roots of a fern and ensured health and strength.

Kombucha fermentation is a lactic acid fermentation process. There are two lactic acids, one being 'L(+) lactic acid' and the other 'D(−) lactic acid'. The (+) lactic acid in Kombucha detoxifies, cleanses, purifies and strengthens the body.

The metabolic by-products of the fermentation produced by the tea and sugar include gluconic and glucuronic acids, acetic acid, carbonic acid and vitamins B1, B2, B3, B6, B12, folic acid and various enzymes. Usinic acid produced is antibacterial and partly antiviral. Acidic acid bacteria produced are antagonistic to streptococci, diplococci, flexner and shigella rods *(Nosredna)*.

Part IV:
MAKING THE BEVERAGE MORE INTERESTING

KOMBUCHA – COMMERCIAL PRODUCTS

Kombucha beverages are already commercially available. Some people don't want to brew the liquid themselves and therefore prefer to buy it. To find a good commercial beverage is not easy. Some beverages show all the components of a good product on their label. The contents, however, may not be quite so promising.

With a commercial Kombucha beverage, we are dealing with a fully active food beverage. The fermentation process is still working fully and the taste will change daily. It is possible that bottles purchased commercially will have different tastes, even though they have identical labels. Depending upon how long it has been stored and at what temperature, the taste may change and sour. People who purchase a bottle once, to try, can be put off forever by a sour beverage. They will never discover that Kombucha is a delicious product. A change of taste, due to storing, is a sign that no preservatives have been used and that the beverage is a complete living food item. With preservatives, we usually think of the chemical additives which are mentioned on the label.

Heating the drink will prevent further fermentation and should not be used with Kombucha. Microwave oven treatment will destroy the micro-biological components of the beverage (which are the main sources for its healing effects). Commercial beverages are almost always manufactured with green or black tea. I am only aware of one manufacturer who incorporates herbal teas with a mixture of green tea. Herbal teas produce the beverage just as quickly and will increase its shelf life without destroying the micro-biological balance. At the same time, the healing effect of the herbs is also incorporated into the beverage. This unique method has been developed with finesse by Mr Jose Perko with years of experimenting and experience. The Papaya 5 Range is a Pawpaw leaf extract concentrate with a basis of Kombucha and consists of:

Papaya 5

Celery and Papaya 5

Chilli and Papaya 5

Beetroot - Dandelion and Papaya 5

Guava Leaves and Papaya 5

Celery is used as a natural diuretic and assists the relief of rheumatism. Chilli (Cayenne) is used to stimulate and strengthen the circulatory and cardiomuscular system. Beetroot and Dandelion are used as blood purifiers and to support and strengthen the liver. Guava leaves have a high content of vitamin C, and are used in South-East Asia as a natural analgesic and to control diarrhoea.

Papaya 50 and Kombucha Concentrate is the extract of the Kombucha culture itself, combined with a super-concentrated extract of pawpaw leaves (10 times stronger then the Papaya 5 range).

A commercial culture produced in Australia.

Kombucha and pawpaw products from The Way of Life Sanctuary, *Queensland, Australia.*

HOME-MADE; THE MOST IMPORTANT STEP FOR THE HEALING PROCESS

Natural healing does not depend on the use of medicines alone to cure the symptoms. Natural healing is healing yourself. It takes place when a human being becomes a part of nature and heals in harmony with nature, applying various natural remedies. When Pastor Weidinger kindly agreed that I could quote from his various publications, he made the following interesting remark: *"What I would like to especially highlight is the fact that I do not regard Kombucha as a trade product, but a home-made one. I support self-made Kombucha instead of commercial processing. Because of this, I generally have to disallow the use of my comments because I regard the fact that someone who makes the effort to obtain information is already on the way to healing."*

KOMBUCHA - MIXED BEVERAGES

Kombucha can be mixed together with other beverages. Children, for example, love it with apple juice. Kombucha fermented with raspberry leaves tastes very similar to apple juice after three to four days and many children do not even notice that they are drinking Kombucha. In addition, raspberry syrup or blackcurrant syrup is also very popular with little ones. Lemon is popular with young and old alike. Some people are convinced that whisky makes an ideal mixture! When Kombucha became popular in central Europe, people organised parties where the only beverages available were Kombucha mixtures of various types.

KOMBUCHA CHAMPAGNE

When I visited Mr Perko recently in Queensland, I was able to try many different types of Kombucha. It was a similar experience to wine-tasting in Austria or Italy. The last glass I was served was a champagne, a very fine product which I assumed was an expensive treat. I was very surprised when I was told that this also was a

Kombucha beverage. Even though the exact recipe has not been published, Mr Perko revealed that he used sugar, honey, elder-flowers, mulberries and a small amount of herbs. I am grateful that I was allowed to taste this and that I was given a bottle of the magnificent liquid. Perhaps one day the recipe will be made public.

HERBAL TEA REFINEMENT THROUGH KOMBUCHA FERMENTATION

Kombucha has been used by many people over the years as a gourmet product and healing remedy. The different taste com-binations are endless. The main factors for the fermentation are sugar, constant temperature, time of fermentation and time of storage. With the differing teas that can be used, the varieties of taste are endless. In short, there are many thousands of different Kombucha beverages with different effects, and I could write a large book on all the different healing remedies and tastes pro-pounded.

It is important to remember that the fermentation process of Kombucha will retain the characteristics of the tea used. Accord-ing to some, the fermentation process even highlights some of the characteristics of the teas. When brewing Kombucha, a very light tea is generally used. People who have a nervous reaction to black tea will have this effect increased with Kombucha, even though a lighter tea is used. For them, more beneficial results are usually achieved with herbal tea. It is similar to products such as sau-erkraut or fermented garlic, where the garlic will retain its effects after fermentation but without the smell associated with it.

When brewing Kombucha, however, not all teas are suitable. Volatile oils can harm the fungus and even kill it. The amount of herbal tea used should be twice that of black tea, i.e. approx. 10 grams (¼ ounce) per litre (2 US pints). A heaped teaspoon of leaves is approx. 5 grams.

WHICH HERBS ARE SUITABLE FOR THE KOMBUCHA PROCESS?

Some of the most popular herbal teas are: raspberry leaves, hawthorn, valerian, stinging nettle, dandelion, elderflower, straw-berry leaves, blackberry leaves, etc. If you are at the beginning of your Kombucha brewing career, I advise against using herbal teas that grow close to the ground. Start with teas such as raspberry leaves, elder or hawthorn. Ground-loving teas are susceptible to larger amounts of bacteria and germs which may be neutralised to some extent but may still cause the fungus to grow mould. Al-though yarrow has a large amount of volatile oils, it can still be used in a mixture of about 20%. A mixture of yarrow, dandelion, stinging nettle, elder and raspberry leaves in equal parts is a very healthy mixture, which also has an excellent taste.

One thing needs to be stressed. The quality and healing extent of natural medicine depends upon the components, how the components have been processed, and how fresh they are. We need a very healthy functioning fungus culture in order to produce a good herbal tea. I have farmed herbs for many years and have always stressed that the best herbs are those harvested from one's own garden. All herbs, with few exceptions, need to be fresh. A general rule is that once you have your new harvest, the old belongs on the compost heap. People who wish to have good results need to know when the herbs were harvested. Roots and barks generally have a longer shelf life.

Other important factors relate to the time of the harvest and the way in which the herbs have been dried and stored. Some herbs require harvesting in the morning dew and others in the afternoon sun. Most roots should be harvested during rest periods of the plants, others when the plant is in blossom. The drying process has to be quick to avoid mould growing which, in turn, could harm the Kombucha fungus at a later stage. Herbs should not be dried at too high a temperature. Direct sunlight should also be avoided during the drying process. People who compromise with the manufacture of a health remedy cannot be sure that it will be effective. In 1984 I produced a video, entitled 'Back to nature with medicinal herbs', which covers the basic requirements for obtaining the correct herbs.

Most herb books on the market today are based on the experiences of our ancestors. We have inherited our knowledge of health from a time when no pills were produced and when research and industry did not attempt to obtain active ingredients, extract them and sell them in marketable form. Knowledge of healing has been built on successful experiences of working with nature. When the balance of nature is disturbed, we have to have experiences spread over long periods of time before we can be certain of the effects. Will new experiences with an unbalanced nature be a successful experience?

For Kombucha brewing, the recommended mixture of tea with various herbal teas is given below, with the ratio of ingredients given in brackets. For example (1/4) means that one part of this herb tea should be used to four parts of your standard black or green tea mixture. Some common herbal teas and their applications are listed as follows:

AGRIMONY *(Agrimonia eupatoria)*: A bitter tonic (strengthening). Astringent to the gastro-intestinal tract. Useful in the treatment of diarrhoea. A good remedy for liver complaints. Effective for coughs, sore throat, bronchitis (gargle); strengthens the immune response. (1/3)

ANGELICA ROOT SEED *(Archangelica)* is one of the main general tonics: stimulant, stomachic, carminative. (1/4)

BARBERRY *(Berberis vulgaris)*: A good remedy for stomach, liver and kidney disorders, chronic constipation and jaundice - has diuretic properties. (1/4)

BEDSTRAW *(Gallium verum)* cleanses kidneys, liver, pancreas and spleen; supports the lymphatic system and the treatment of tumours. (1/3)

BILBERRY LEAVES *(Vaccinum Myrtillus)* must be taken with

great caution and for a short period of time only, followed by an equally long break, also used as infusion. (1/3)

BUCKTHORN *(Rhamnus Frangulata)*: A mild laxative, cathartic. Can trigger menstruation. (1/4)

CALENDULA *(Calendula officonalis)*: To tonify blood vessels as in varicose veins and for infections especially in the lymphatic system. Regulates menstruation; blood cleansing and a mild laxative. (1/3)

CHAMOMILE *(Matricaria chamomilla)*: Tonic carminative and sedative. Useful in neuralgic pain, infantile convulsions, stomach and gastro-intestinal disorders. (1/4)

COLTSFOOT *(Tussilago farfara)*: Expectorant for cases of coughs and asthma. (1/3)

DANDELION *(Taraxacum officinalis)*: For the stimulation of the function of liver and gall bladder. It is used in all digestive disorders and in rheumatism and gout. (1/3)

ELDERFLOWER *(Sambucus nigra)*: A splendid spring medicine, blood purifier, gentle laxative. (1/1)

FENNEL SEED *(Foeniculum vulgare)* is often used for babies suffering from dyspepsia, diarrhoea and flatulence. Mixed with other appropriate herbs to help relieve complaints of the respiratory and digestive tracts. Used for bronchial catarrhs and other severe respiratory infections and colic (1/6).

GENTIAN *(Gentiana lutea)*: A time-honoured tonic, promotes digestion.

GOLDEN ROD *(Solidago virgaurea)* is very beneficial for kidneys and urinary tract. An excellent diuretic for chronic infections of the kidneys, as it increases the flow of urine. Stimulates the blood supply to the kidneys and veins of the legs. (1/2)

HAWTHORN *(Crataegus oxyacantha)* is the cardiac tonic. It stimulates the supply of oxygen to the heart muscle. Regulates blood pressure and heart action. (1/2)

HORSETAIL *(Equisetum arvense)*: For all disturbances of the urinary tract, also for bed-wetting. (1/2)

MARSHMALLOW ROOT *(Althaea officinalis)*: For inflammation of the whole digestive tract and of the respiratory organs etc. (1/3)

MOTHERWORT *(Leonurus Cardiaca)*: An antispasmodic heart tonic for anxiety; allaying nervous irritability due to overactivity of the thyroid glands. Excellent tonic for female complaints and nursing mothers. (1/3)

MULLEIN *(Verbascum densiflorum)* is excellent for most lung complaints including coughs, bronchitis, hoarseness of the throat. Relieves irritation of the respiratory mucous membrane. It has a cleansing and expectorating effect on the mucous in the respiratory tract. (1/2)

OAK BARK *(Quercus robur)*: For haemorrhoids, leg ulcers and chilblains. Foot perspiration. (1/3)

PEPPERMINT *(Mentha pip.)* is best known for curing stomach and intestinal complaints. It is slightly disinfectant, spasmolytic, carminative, promotes the production and flow of bile - helps flatulence, vomiting, soothes nausea. Indications include stomach upsets, poor digestion, stomach cramps, diarrhoea, nervousness, restlessness, colic. (1/5)

PROSTAWORT *(Epilobium roseum)*: The best herb for all kinds of prostate, bladder and kidney disorders. (1/2)

RASPBERRY LEAVES *(Rubus idaeus)* have been useful with diarrhoea and stomatitis, also been used as a mouthwash. (1/1)

RIBWORT PLANTAIN *(Plantago lanceolata)*: Effective in the treatment of infections of the respiratory tract; catarrh. Removes secretions from the bronchial tubes. Tonifying and stabilising for tissue because of its high content of silica. Has been used as a remedy for tuberculosis. External use for wounds. (1/2)

SAGE *(Salvia officinalis)* is a blood cleanser, suppresses perspiration. In malfunction of the thyroid glands or during menopause it assists poor, irregular and painful menstruation. Gargles are the best remedies for laryngitis, tonsillitis, ulcerated and sore throats.(1/4)

SHEPHERD'S PURSE *(Capsella bursa pastoris)* is a very important homeostatic plant. Especially uterine haemorrhages, menstrual difficulties, kidneys, nose, mouth, bladder and respiratory tract. Regulates high or low blood pressure. (1/3)

SPEEDWELL *(Veronica officinalis)* is a blood cleanser, regulating menstruation, perspiration and mucous discharges. Tonic and soothing effect on nerves.

ST.JOHN'S WORT *(Hypericum perforatum)*: Useful in nervous disorders, depression and hormonal imbalance. External oil applications are effective in the treatment of wounds, burns, frostbite, rheumatism and backaches. (1/4)

ST.BENEDICTS *(Geum urbanum)*: For febrile conditions, asthma and obesity. It stimulates the digestion and strengthens the body. (1/3)

STINGING NETTLE *(Urtica dioica)*: For urinary and skin disorders, as well as for all rheumatic complaints - to reduce excess acidity. (1/3)

VALERIAN ROOT *(Valeriana officinalis)* is very popular for use in nervous conditions and insomnia as a calmative. (1/5)

WHITE DEAD NETTLE *(Lamium album)* is effective in menstrual difficulties and vaginal discharge, external and internal applications. A uterine tonic – also for the treatment of the urinary tract. (1/2)

WILLOW BARK *(Salix alba)*: Reduces fever and alleviates pain. It is a very effective remedy in the treatment of rheumatism, inflammations, internal bleeding and makes a good diuretic. It can be used as a gargle for tonsils and externally for wounds and burns. (1/3)

YARROW *(Achillea millefolium)*: For bleeding, haemorrhoids, varicose and menstrual disorders. It improves the circulation and is antispasmodic. (1/4)

There are many slimming, sleeping, bladder, asthmal and other mixtures one can develop with different teas for all complaints. It is advisable to alter mixtures after a few weeks to give the body a new stimulant for a better effect as well for avoiding negative side effects, but also consider:

Is it for general well being? Then the best taste is important. Green or black tea is mixed with all sorts of teas for refreshment and taste, such as wild blackcurrant, raspberry tea (fruit and/or leaves), lemon tea, tropical fruit teas and so on.

Is the tea beneficial for you? For example, when you are suffering from constipation, peppermint is not the right tea. Green and black tea can affect your blood pressure. If you are not feeling well, use a different mixture!

Is it to treat a specific complaint? All the teas used should be beneficial for the illness to be treated. The numbers listed below refer to stand for parts used (ie. one teaspoon of each). For instance:

Asthma: *Mixture A:* St. Johns wort (1), St. Benedicts root (1), Valarian root (1), Angelica root (1), Melissa (2).
Mixture B: Thymus (1), St. Benedicts root (1), Elderflowers (1/2), Marshmallow root (1), Melissa (1).
Mixture C: Mullein (2), St. Benedicts root (1), Ribwort Plantain (1).

Bladder complaints: Horsetail (1), Calendula (1), Prostawort (1).

Insomnia: Hops (1), St. Johns wort (1/2), Valarian root (1/2), Melissa (1), Passion flower (1/2).

Slimming: Lady's Mantle (1), Angelica seed or root (1), Calendula (1/2), Corn silk (2), St. Benedicts root (1/2), Buckthorn bark (1).

WHICH IS THE BETTER? GREEN OR BLACK TEA?

I am often asked which is the best tea to use for a Kombucha fermentation. In the Eastern countries of origin only green tea was used. In Russia and the German-speaking countries, mainly Russian tea – black tea – is used. Tea is the most popular drink in the world – not beer or Coke as some people want to believe. Besides green and black tea, there is the semi-fermented Oolong tea. The source of all common teas on the market is the tea bush *Camellia sinensis*.

The teas known as English breakfast teas, Russian teas, etc. are black fermented teas. Some of these are blended with other teas, e.g. with other herbs like lavender or with Bergamot which gives 'Earl Grey' its flavour. Most people in the Western world drink black or Oolong tea rather than green tea, probably because of the stronger flavour. Many individuals feel a stimulating effect when drinking black tea, similar to that experienced from coffee. Green tea, on the other hand, doesn't stimulate as much, but has a steady and longer-lasting effect.

Alkaloid caffeine is responsible for the stimulating effect of tea which has between 1% - 5% caffeine. Recently, green tea has been frequently mentioned as a potential preventative of cancer. Dr Robert E.Willner (USA), author of the book *The Cancer Solution* mentions 'delicate pale tea' as a source of anti-carcinogenic substances. Green tea contains the chemical *Epigallocatechin Gallate* (EGCG), which inhibits the growth of cancer and lowers cholesterol.

'Green tea gives health and longevity' is an old Buddhist saying. The Japanese extend the healing benefits of green tea by eating the leaves since, when tea is drunk, a considerable portion of the precious vitamins and trace elements are thrown away. Professor Kazutami Kuwano of the Kasei Gakuin Junior College, (Chikaro Shimoaka *Green tea – more than a health drink*) closely investigated the 'tea-food'. He found that the vitamins A and E which are contained in green tea, are not absorbed by the body since these vitamins are not water-soluble.

Calcium, iron and vitamin C are only half utilised in the tea when drunk, compared with that absorbed from the tea when eaten. To make three cups of green tea one uses approximately six grams of tea leaves – a teaspoonful. This quantity of tea when eaten contains 50% of our daily needs of vitamin E, and 20% of our vitamin A requirements. There are virtually no calories in tea leaves. The healing effect of green tea is determined by the quality of the product. This depends on where the tea is grown, the harvest (vital young leaves or old leaves) and the processing procedures. Loss of quality of tea or of any other medicinal herb can also occur with poor conditions or too long storage.

KOMBUCHA'S PLACE IN MEDICAL SCIENCE

There is no doubt that it is better to prevent illness than cure it. However, this is where problems may start for many people. From fear of illness, people use strong natural medicines to prevent disease occurring. This overlooks the fact that these can also have negative side effects.

A typical example is comfrey *(Symphytum officinale)*. This powerful remedy to heal a bone break should be used carefully. Internal use as a prevention can be harmful. Chamomile and peppermint are both valuable herbs when used correctly, but both can have negative side effects if used wrongly and many herbs can be harmful if used for long periods of time. Only a few months ago, I was asked by a woman what could be done for constipation. As it turned out, the woman had consumed peppermint and chamomile tea daily. She stopped drinking the teas, and the problem was solved.

Illness can be prevented by a balanced life. Food is one of the three main components needed for healthy living. As a known aid to digestion, Kombucha can become an important part of this. In this respect, Kombucha acts as a general strengthening remedy by enhancing the body's defences. Kombucha is therefore an important part in a preventative diet.

In our technical age we need to prove everything scientifically. When one is not able to do so it is not taken seriously. 'Where is the scientific proof' is a common question when talking about Kombucha and similar subjects. When I look at my books about agriculture, which are three decades old I shake my head and cannot help but question the value of scientific proof. These books are based on scientific studies and now we have new research that shows the previous research was wrong.

The problem that modern science has with Kombucha is that it is nearly free (sugar and tea only) and that everyone can produce it at home in the kitchen. Scientific tests have been done in some countries, but only as to how people react, not why they reacted in this way. Powerful chemical companies are **not** interested in Kombucha since on the one hand everyone can produce this delicious low cost medicine at home and on the other Kombucha works only as it is, in its natural balanced condition. The modern way of extracting the active ingredient and to produce in this way a marketable medicine is not suitable for Kombucha. Nobody will fund the scientific research because there will be no product to sell. We may just have to wait for scientific proof and be happy with the results millions of people are experiencing.

Part V:
THE THERAPEUTIC VALUE OF KOMBUCHA

MODERN MEDICINE AND KOMBUCHA

Kombucha results cannot be fully explained scientifically, but experience and anecdotal evidence have proved its efficacy over many centuries. In scientific trials only green or black tea were used. These teas show totally different results after fourteen days of fermentation compared with herbal teas.

The German *'Lebendsmittelrundschau'* (1987) published the following results:

Tea	Lactic acid	Glucon acid	Acetic acid	Ethanol
Common tea	2.94%	2.52%	0.08%	1.07%
Peppermint tea	0.14%	0.04%	0.01%	0.01%
Linden flower	0.007%	0.06%	0.30%	0.04%

The lactic and glucon acids are the two most important components scientists have concentrated upon. With the lactic acid, scientists have researched the 'L(+) - lactic acid'. There are numerous scientific reports: Dr S Hermann and his team, conducting trials in Prague, came to the conclusion that the glucon acid in Kombucha could dissolve gallstones. Dr Wiesner's research showed that Kombucha compared favourably with an Interferon product in gallstone treatment.

With asthma, Kombucha was even more successful than the applied Interferon product. In his report, Dr Wiesner suggested that Kombucha may stimulate the body's resistance to illness, thus assisting the healing process. It may further be regarded as an important food item in its ability to slow virus attack without having side effects. It should also be mentioned that Dr Reinhold Wiesner developed a blood test which electronically monitored energy levels in the blood in conjunction with a medicine being used. In the clinic at Omsk in Russia the following Kombucha trials were conducted:

Dysentery in infants was being treated with Kombucha. After only a few days, it was recorded that the infants were gaining weight again. The diarrhoea became less. After one week of treatment, no further dysentery bacteria could be found in the faeces.

High blood pressure and arteriosclerosis clearly showed better results following 2-3 weeks treatment with Kombucha. Cholesterol levels sank remarkably at the same time.

Inflammation of the tonsils was treated with remarkable success. Patients used Kombucha tea to rinse their mouths ten times a day; the liquid was held in the mouth for fifteen minutes.

COMPARISON BETWEEN IMMUNE STRENGTHENING INTERFERON AND KOMBUCHA

In 1987, the physician and biologist Dr Reinhold Wiesner from Bremen, Germany, published the results of a study dealing with the effects of Kombucha. He tested the effects of the beverage on 246 patients and compared the effects with those obtained from an Interferon product, synthesised from the anti-viral agent that our bodies produce naturally. The patients had illnesses such as kidney disorder, inflammation of the liver, rheumatism, multiple sclerosis and asthma. Kombucha achieved better results than Interferon (203 compared with 183) for asthma, whereas with the other illnesses, the results were only marginally below those of Interferon: rheumatism 92% as effective, kidney disorders 89%, inflammation of the liver 81%, and multiple sclerosis 80%.

CAN KOMBUCHA HEAL CANCER?

Kombucha cannot be regarded as a single healing remedy. Within biological cancer therapy, however, Kombucha is well thought of in the German-speaking countries. Cancer experts, such as Dr Johannes Kuhl, Dr Rudolf Sklenar and Dr Veronika Carstens, have become famous due to their successful cancer therapy where Kombucha plays a major role in revitalising the bowel flora of patients. Dr Reinhold Wiesner, a physician/biologist from Schwanemünde, in Northern Germany, published a very interesting research paper. The anti-viral medicine was tested in comparison to Kombucha. Results showed that when Kombucha was used, the effectiveness achieved was just marginally below that achieved by the other product. Details of the type of tea used were not given. Generally, however, black tea is used in Germany. In Asia and Japan, green tea or Kombu tea is generally used. For cancer treatment, pawpaw leaves and/or calendula *('calendula officinalis')* are currently being used.

Mrs R.E.H wrote me the following letter: *"My 58 year old sister who is a smoker, two years after a right mastectomy, has now got secondary cancer in her ribs, skull and a growth in the lung. Chemotherapy was prescribed. Two weeks after the first chemo, she suffered complete hair loss and excruciating pain in her ribs, requiring large doses of morphine painkiller. She started drinking Manchurian Tea (Kombucha) about the time of the hair loss. Within two days, she discontinued the morphine as the pain had disappeared. Insomnia was no more. The hair on her head started to grow again and didn't fall out while chemo was continuing, four sessions in all – four weeks apart. She could still work full time, except for one day following each chemo session. The bone cancer medication was supposed to swell her up, and weight gain was supposed to be anything up to 5-6 stone! But she gained no weight, although her appetite was hearty. The growth in the lung shrunk, and the specialist said to her 'whatever you are doing, keep doing it'. Of twenty-nine fellow chemo patients, she is the only one still maintaining her job. All others had to leave their place of employment, because they were too ill. She didn't experience the side effects of chemo to such a severe degree as all others did. The scanning of bone cancer has not been done again so far."*

PAWPAW AS CANCER MEDICINE?

Pawpaw was known to me only three years ago as a delicious tropical fruit which, with lemon juice, was good for the digestion. I first heard of its medicinal qualities from a woman writing to the German newspaper in Australia *'Die Woche'* describing how pawpaw had cured her cancer. Mrs H had lost any hope of being healed, when she was made aware of the potential healing in pawpaw leaves.

A more detailed report in the *'Weekend Bulletin'* (Gold Coast, Australia) deals with the healing of Mrs K's cancer of the bladder. Mrs K (a 74-year-old woman) had an operation, but the cancer could not be completely removed and she was supposed to have further treatment in Brisbane. Over a period of three months she used pawpaw leaves. When she ran out of them, she used the skin of the fruit which she boiled. When she went back to the doctor for a further check-up after that period, the cancer had been healed. A check-up four months later confirmed this and Mrs K, who now feels completely whole, is quite convinced her cure was due to the pawpaw.

On the other side of the world, the American scientist, Dr Jerry McLaughlin of the University of Purdue, has also used pawpaw in the fight against cancer. According to him, he has found a chemical component in the pawpaw tree that is 'one million times stronger than the strongest anti-cancer medicine'. There are many reports that cancer sufferers have been healed by drinking pawpaw leaf concentrate.

A Kombucha-fermented pawpaw leaf extract is available in some health shops. Kombucha concentrate is now manufactured in Australia. I have found a number of practitioners who have used pawpaw successfully but who are reluctant to prejudice their work by saying so, for fear of legal implications. The first reports of pawpaw's healing abilities were made in 1978 and resulted in a demand for and subsequent scarcity of pawpaw leaves.

The *'Gold Coast Bulletin'* has published several reports on pawpaw's reputed medicinal qualities, including this one:

"PAWPAW CANCER PLEA BEARS FRUIT. Gold Coast gardeners have responded to an appeal by cancer victims desperate to find supplies of pawpaw leaves. And the Gold Coast man who, 14 years ago, first exposed the leaves as a possible cure for cancer has been tracked down to a Labrador nursing home. The story of how Stan Sheldon cured himself of cancer by drinking the boiled extract of pawpaw leaves was first told in the Gold Coast Bulletin *in 1978. Now research in the United States has given scientific support to his claim, isolating a chemical compound in the` pawpaw tree which is reported to be a million times stronger than the strongest anti-cancer drug.*

"DYING - Mr Sheldon, 88, says the discovery does not surprise him. 'I was dying from cancer in both lungs when it was suggested to me as an old Aboriginal remedy', he said. 'I tried it for two months and then I was required to have a chest x-ray during those compulsory TB checks they used to have. They told me both lungs were clear. I told my specialists and they didn't believe me until they had carried out their own tests. Then they scratched their heads and recommended I carry on drinking the extract I boiled out of the pawpaw leaves.' That was in 1962. The cancer never recurred. Since then, Mr Sheldon has passed the recipe on to other cancer victims. 'Sixteen of them were cured', he said. Mr Sheldon's recipe involves boiling and simmering fresh pawpaw leaves and stems in a pan for two hours before draining and bottling the extract. He said the mixture could be kept in a refrigerator, though it may ferment after three or four days. He said there were times when he found it difficult to obtain supplies of pawpaw leaves. 'Not everyone likes you removing them. They are afraid you will ruin the tree', he said. But Gold Coast pawpaw growers have responded generously to a plea for help last week by cancer victims desperate to find a cancer remedy.

"Home gardeners, a nursery operator and a tourist attraction owner have offered leaves from their pawpaw trees to people seeking a supply. One man, Vern Forrest of Burleigh Heads is a veritable Johnny Pawpaw seed. He has been growing pawpaws and giving away the leaves to cancer victims ever since he read the

Bulletins *original 1978 story about Mr Sheldon. 'I have no doubt that it works', he said. 'I know of people walking around now who should have been dead according to their original cancer diagnosis. But the pawpaw treatment helped them to beat the cancer.' Betty Ellingsworth, who manages a Nerang nursery, said cancer victims were welcome to the leaves from the nursery's four pawpaw trees. And Pat Washington has offered leaves from trees growing at her Tallai Hills Home.*

"SUPPORT - Bob Brinsmead, who operates the Avocado-land tourist attraction on the Tweed, also has pawpaw trees which he will make accessible to cancer victims. 'I am not surprised by the American research', he said. 'There is a pawpaw ointment made in Brisbane which has been very successful in treating skin lesions. And I once painted the sap from a pawpaw stem directly on to a worrying growth behind my ear. It got rid of it.' Dr Jerry McLaughlin, the Purdue University scientist who discovered the anti-cancer substance in pawpaw, has become a victim of publicity. Attempts to contact him at the university's pharmacy school in Indiana are met by a taped message in which he says it is 'impossible for him to respond to all the inquiries that have followed the reports of his research, which is 'still in the stage of animal testing."

New on the market is a Kombucha-Pawpaw concentrate. Mention should also be made that the skin of the pawpaw fruit *('Cribb')* can help with sores.

EARLY DETECTION OF CANCER INCREASES HEALING SUCCESS

The earlier cancer can be diagnosed, the better the chance of healing. Unfortunately, with the present standard of medical diagnostics, cancer is often already in an advanced state when discovered. According to Dr Sklenar, prior to the acute start of cancer, illnesses can be detected which may have been there over a long period of time. Dr Sklenar developed an early detection blood test with which he claims to be able to detect cancer early through his 'blood colour' method. In his search for early detection, Dr Sklenar also adopted iris diagnosis which he found of valuable help. In the iris he found brown to black sediments which led him to the conclusion that the colon was not functioning fully. Patients in a pre-cancer state often have digestive problems. Even though diagnosis of the iris is only successful in 80% of cases, it is still a precious tool in the early detection of the disease.

Other uncommon diagnosing methods also appear to be useful tools in the early detection of cancer. Here's one example: I was of the opinion that one of my close friends had an acute problem of the liver. I consulted a practitioner without telling my friend of my concerns; the practitioner concluded, using well-developed intuition, that the person I was describing might have cancer of the liver. Tests conducted in famous hospitals in Germany and Switzerland had not confirmed cancer. The patient had had frequent check-ups for other illnesses in hospitals. Only 3 years after the health practitioner had suggested cancer of the liver, my friend was diagnosed with just that, which resulted in death three months later. The questions remains whether the patient might have been able to alter the outcome by changing her diet, quitting smoking and altering her bed position.

AIDS

Two of the benefits from drinking Kombucha most often mentioned are the energy it releases and its immune-system boosting effects. I often thought that Kombucha might benefit AIDS sufferers but, initially, did not have any reports of this ancient remedy being used to treat this modern disease.

Recently, however, there have been widespread reports that Kombucha use in HIV+ communities is growing fast. It is being said that the Kombucha tonic can reverse the symptoms of AIDS. In the USA, which has an estimated one million HIV+ infected people, the news of a possible miracle cure is spreading rapidly. In some publications it is being claimed that components in Kombucha partly 'inactivate the viruses'; that high p24 antigen levels can be reversed, a rise in T-cells achieved and weight increases of 5 - 7 kg recorded. All these are measurements of improvement. Long term AIDS survivors also report that Kombucha has a positive result.

The 'blob', as Kombucha is sometimes called, is passed on to friends as a remedy for AIDS with the additional advantages of its palatability, its apparent safety and low cost to commend it further. *"('Positive Living', Los Angeles, USA; 'With Compliments', Sydney, Australia; 'Whole Life Times', Malibu, California)"*

"Kombucha might have remained within the confines of the holistic community were it not for the August 1994 issue of 'Positive Living', a newsletter published by AIDS Project Los Angeles (APLA). Under a headline claiming 'Reborn on the 4th of July' was the photo and story of Joe Lustig, a Long Beach man who has AIDS. As that month's subject of a regular profile on long-term AIDS survivors, Lusting told Paul Serchia, the newsletter's editor, of his 'startling breakthrough' on Independence Day.

"I spent more of 1993 in bed', Lustig says. *'I could hardly get to the bathroom without a walker.'*

"Lustig first tested positive in 1985 and was diagnosed with AIDS in 1990, after coming down with pneumocytis carinii pneumonia (PCP). He developed a litany of increasingly dangerous infections: meningitis, encephalitis, pancreatitis, all complicated by dementia. *'I was in such a fog I couldn't pay my bills. I had this constant underlying feeling of tiredness and sickness. A very heavy, toxic feeling. It was there every day when I got up, and every night when I went to bed.'*

"After less than one week on Kombucha, however, Lustig says he felt better. *'I have to emphasise that life was not worth living any more. Then I woke up one morning and my first thought was that I had died and gone to heaven. Everything felt better.'*

"According to Lustig, he got out of bed and did something he had not done in four years. He went for a run. He also insists that no other changes in his medications took place that might have accounted for the change.

"Lustig rides his bike three or five miles a day now, reports increased energy, longer waking hours, and better cognitive function. He has also given up one of his medications. *'I haven't taken Nizoral* (a drug that fights thrush) *for ten weeks'*, he says, the length of time he has been drinking the tonic." *(New Age Journal, November 1994.)*

R.D. from B.B. wrote: *"I was introduced to Kombucha by an American friend who brought a 'mother' fungus from the U.S.A. This person has been HIV+ for the past eleven years. He was quite exited about Kombucha and has been drinking it for nine months now, mainly to keep his T-cell count adequate. He is healthy and robust looking, maintaining a positive attitude, moderate lifestyle and good diet. My friend says that since he has been drinking the tea fungus, he feels more energetic and incredibly well. The drinking of Kombucha tea has swept the world of HIV diagnosed people and AIDS sufferers in America over the last year or so, as it seems to be a very good natural aid for them".*

Another correspondent writes: *"I started drinking Kombucha about three months ago and find it most helpful. About five years ago I was diagnosed HIV+. My T4 count then was 660 and had been dropping over the years to 390 before starting your tea. Within 2 months my count was up to 570 – it has not been that high for nearly two and a half years. I also have found it helpful with my arthritis and sinus problems."*

KOMBUCHA, HEALING REMEDY FOR ASTHMATICS

Asthma is a disease spreading at an alarming rate, triggered by many differing environmental factors and combinations. Kombucha has a positive effect upon digestion by detoxifying the body. This detoxification has a positive effect on asthma. Especially interesting is a trial result of Dr A Wiesner. He tested the effect of Kombucha on 246 patients in comparison to the immune-strengthening product Interferon. Generally speaking, asthma sufferers reported a considerable improvement when using Kombucha. Several herbal mixtures have also had a positive effect with asthmatics. It is more desirable, however, to find the cause of the illness so that asthma can be totally avoided.

SLEEPING DISORDERS

Sound sleep is just as important as eating and drinking. We can survive several weeks without food, but not without sleep. Each individual must determine how much sleep he or she requires. Some require a mere five hours of sleep per day, others need ten. Stress and unsolved problems can be the cause of sleeping disorders along with environmental factors. People having problems sleeping through the night have found that a glass of Kombucha is helpful.

PROSTATE DISORDERS

One of the principal problems associated with male ageing is prostate disorder. Approximately one third of men have prostate surgery and a further one third experience problems of one sort or another. The tea of the alpine willow-herb *('epilobium roseum* and *epilobium parviflorum')* has been used for many generations in the Alp regions of Europe to counter problems associated with the prostate. This remedy has been available in Australia since 1983, and sold as 'Prostawort'. When combining this with a Kombucha fermentation it should be mixed in a ratio of about a fifth (20%) to green tea. The tea should not be more than twelve months old. With inflammation of the bladder this tea has shown some good effects. A mixture of Horsetail, Calendula and Prostawort in equal parts has also shown good results. An elderly lady suffering from bladder inflammation for decades now drinks this tea mixture at the first sign of the problem recurring, eliminating the problem before it sets in.

DIABETICS AND KOMBUCHA

In many parts of the world, the Kombucha culture is being passed on to friends in the reassuring knowledge that there are no negative side effects. The question is often raised, however, about the possible effect of the sugar on diabetics. In fact, the fermentation process converts sugar into other products after some 6-9 days, (see graph p.38). Therefore, as long as the brew is not drunk before the fermentation is complete, there should be no problem for diabetics. Pastor Weidinger knew this from first-hand experience as a diabetic himself.

Should diabetics use caution with Kombucha beverages that are being sold in stores? Commercial production may generate ambitious advertising in which all claims may not be met. Where these beverages contain unfermented sugar they must be avoided by diabetics. Some herbal teas when mixed with green teas are suitable for diabetics; these are Billberry leaf tea *(Vaccinium myrtillus L)*, Knotgrass *(Polygonum aviculare L.)*, Burdock *(Arctium lappa L.)* and Chicory *(Cichorium intybus L.)*.

DOES KOMBUCHA DISSOLVE BLADDER STONES?

Dr Herrmann researched whether Kombuchal (Kombucha extract) could dissolve bladder stones. Phosphate stones were inserted into the bladders of male rabbits. Over several weeks, the animals received Kombuchal twice or three times daily, and the phosphate stones slowing reduced in size, eventually being discharged through the bladder. An increase in calcium and phosphate could be recorded in the urine prior to the stones disappearing. The Czech scientist discovered that phosphate stones could be largely dissolved in the laboratory with glucon acid. He believes that glucon acid, taken daily over a period of years, can be absorbed by the human organism. Free glucon acid, chemical or biological, can be produced with a Kombucha brew based upon green tea and black tea.

MULTIPLE SCLEROSIS

In trials, Dr A Wiesner tested the result of Kombucha in comparison to the immune-strengthening medicine Interferon. Kombucha showed an 80% effect in comparison to the modern drug. In a letter published in the magazine *Op Zoek*, Mrs M.W. from Holland, who suffers from multiple sclerosis, wrote of her experience of taking Kombucha. In April, 1989, she started to drink Kombucha and experienced complete detoxification. She is no longer tired and has also got her regular driver's licence back. She writes in her letter, dated February 1990, that, after having taken Kombucha for half a year, she even plans to go skiing again. She wants other people to become aware of the benefits of the Kombucha fungus so that they may experience the same improvements.

DIARRHOEA

Kombucha can be used successfully by people suffering from diarrhoea. Tea mixtures of black and plantain tea are often recommended. Scientific trials have been conducted at Omsk in Russia with infants suffering from bacterial dysentery. Only days after the use of Kombucha drops was introduced, the diarrhoea retreated. The weight of the infants increased and so did their appetites. The dysentery bacteria could not be found after a week.

CHRONIC CONSTIPATION

Remarkable in almost all reports of treating illnesses with Kombucha is the effect that the beverage has on secondary illnesses, such as chronic constipation. This raises the question, once again, which comes first, the chicken or the egg? It was Hippocrates who said that death sits in the bowels and that a bad digestion is the root of all evil. Modern scientists say that the essential requirement for any healing process is a healthy intestinal flora.

Kombucha, with its living bacteria and yeast components, improves and aids digestion. Other remedies for constipation can lead to diarrhoea if used excessively; Kombucha applies a balance. I have, myself, conducted trials by drinking two litres of Kombucha and more over a period of fourteen days and have had good results. One's own experiences are frequently the most convincing. I have always suffered from a nervous stomach and intestinal problems, which is one reason why I have become a so-called Kombucha fanatic. Constipation has long been a problem for me, especially when travelling. In 1982, this led to an operation for intestinal closure. All remedies I had previously tried, both commercial and natural, only had limited success. The dosage either had to be increased after a short period of time or I had to change to another remedy.

With Kombucha, I now seem to have found the one remedy that really helps – and tastes good at the same time. With constipation it is important to investigate the cause; this could be diet but it is also true that not everyone needs to go on a daily basis! Many people tend to panic if they don't have the desired evacuation daily and grab tablets to change this which, in my opinion, is the worst thing anyone can do. Even natural remedies taken on a regular basis can be harmful. Headaches and constipation are generally the most common symptoms experienced when something is wrong with the body.

The cause, in most cases, is our diet and insufficient exercise. In some cases it may be due to stress or to the medicine one is taking. People who often delay going to the toilet because they do not have the time will programme their bowels to become lazy. This can often be solved by merely changing the diet to wholemeal products, vegetables, fresh salads and also natural fat such as butter, cold pressed oil, as well as nuts, etc. Fresh produce should make up at least one third of our diet. One day per week only raw foods should be consumed or a no-food-at-all day (as is practised in many religions) can be made a rule. Drinking a glass of Kombucha in the morning provides a vital aid to the digestive process. Laxatives – including naturally-based laxatives such as senna –

should be taken only as a last resort, when nothing else works.

A mild tea with a laxative effect, like buckthorn, can be used. However women should use caution when using this tea as it may lead to menstrual problems. Those using laxative teas should stick to the milder types and change often to avoid any long term ill-effects. The same goes for tea mixtures. Plum juice and dried plums should also only be consumed for a short period of time.

J.A. from North Queensland wrote: *"I have recently been introduced to the miraculous Kombucha and I am ecstatic with results to date with regard to constipation and stomach pain, which no longer haunt my days."*

PSORIASIS

Kombucha is recommended by doctors who have acknowledged this fermented beverage as a help for psoriasis. Ms S. from B: *"I have been drinking the tea* (Kombucha) *for a few weeks on and off, I have psoriasis on parts of my body. The 3 inch patch I had on my elbow has now diminished to half its size."*

A woman who suffered from the disease for well over thirty years, during which no other treatment was found to be of help, regained soft smooth skin again after drinking Kombucha.

STOMACH AND BOWEL DISORDERS

The use of Kombucha in conjunction with other treatments has had, time and time again, the side effect of curing stomach and intestinal problems. One might conclude that, because the intestinal flora was reinstated through drinking Kombucha, the cause of the original illness could have been disturbed digestion. As with all uses of Kombucha, the first positive sign is usually a balanced digestive system. The rebalancing of intestinal flora is not only to do with the healing of stomach and intestinal disorders, but is the key to healing and general wellbeing.

FLUID IN THE LEGS

Several reports indicate that fluid in the legs will decrease with the help of Kombucha. My father (then 87 years old) had to visit the doctor at least twice a week to have the fluid removed from his knees. Through drinking stinging nettle tea he did not require any treatment for the following eight years.

A woman in her 50s in England, who suffered regularly from swollen ankles (oedema) found that within two months of drinking Kombucha she no longer had the problem, and it has not recurred.

CHOLESTEROL

The media have released many reports on cholesterol, although scientists are not all in agreement about it. High blood pressure, high cholesterol and the resulting blockage of arteries, are regarded as the main cause of heart attacks. Dr Herrmann, with his team of scientists in Prague, conducted animal trials, in which he increased cholesterol levels up to thirteen times above normal. The trials were conducted with the product VigantolR. Dr Herrmann fed cats the deadly dose and, at the same time, gave them Kombucha drops. The result was that the cats were able to survive this dose. He reported that the combination of VigantolR and Kombucha did not lead to an increase in cholesterol levels. Although animal trials cannot always be related to human beings, this test would tend to confirm what people using Kombucha have found themselves. In the Russian clinic of Omsk, a reduction of cholesterol levels was discovered after using Kombucha for a period of three weeks.

HIGH BLOOD PRESSURE

High blood pressure and associated illnesses are at the top of the sickness lists in Western society today. Kombucha is often quoted as a remedy against high blood pressure. One of my friends who suffers from this ailment tried it and some improvement was noted, but fresh hawthorn tea had better results than that achieved by Kombucha on its own. I recommended fermenting Kombucha with hawthorn tea. Afterwards, the results of blood pressure tests, taken both at home and by the doctor, were that the blood pressure had become normal. My friend M.W. commented that it seemed that neither Kombucha tea nor hawthorn tea taken on their own achieved such good results as that which Kombucha fermented with hawthorn tea was able to achieve.

These cases should not be generalised; however a similar observation was made by another high blood-pressure sufferer, Mrs M.Y. who wrote:

"I have been drinking Kombucha for the past four weeks made with hawthorn tea, to bring down my blood pressure. I take approximately 200ml per day along with two cups of freshly brewed hawthorn tea. A cerebration of the two teas has proved extremely effective in keeping my blood pressure within the normal range, and I have been able to dispense with prescribed medication. However, because of a previous health problem, I need to be certain that I am doing the right thing and would appreciate any advice you are able to offer in this regard. In your opinion, is it advisable to take so much hawthorn tea/Kombucha over a prolonged period, or should there be 'rest' periods? If my blood pressure should rise during these rest periods, is there an alternative substance I can use?"

My response is that first of all, it is strongly recommended that you should never dispense with prescribed medication without consulting your practitioner. This may present some difficulty, since some physicians automatically advise against traditional treatments such as herbal medicine. They advise stopping any herbal treatment in favour of the recommended chemical one. Other more open-minded practitioners may advise their patients with more objectivity.

The recommended dose for high blood pressure is two cups per day; for low blood pressure one cup per day. In the case of

Mrs M.Y., I would say that her daily intake of 7oz (200ml) of Hawthorn/Kombucha, plus two cups of hawthorn tea was too high a dosage. 'Rests' are usually advisable, since the body adjusts to treatments. However, hawthorn seems to be an exception. There are many people who have been taking hawthorn on a daily basis over decades with great results. Nevertheless, there are some warnings to be heeded. Hawthorn was banned in Australia for its alkaloid content. Overdoses can have negative side effects. As with other remedies, it is the dose which makes the medicine - or the poison.

My father's experience of blood pressure and heart problems provides a good example of what I mean. Born in 1899, he had to join the army at the young age of 17. The military doctor who examined him advised that he should be discharged after a short time of training owing to his heart condition. During the last twenty years of his life, he took hawthorn tea or tincture regularly to control his heart and his blood pressure. When he was hospitalised at the age of 96 and his death was expected, I asked the doctor how much time he had. The doctor replied that the heart was so strong, it was preventing his death!

Mrs.R.H. wrote: *"I thought you would like to know that people with high blood pressure may have to reduce medication while taking Kombucha"*.

Experience has shown that people using Kombucha with certain medication may experience a fall in blood pressure; if this is likely they should always consult their physician.

CHRONIC FATIGUE SYNDROME (ME)

After the publication of the first edition of this book I received an unusually high response of comments and hints. Many people asked why I hadn't written about the positive effects of Kombucha on Chronic Fatigue Syndrome, since Kombucha is the only medicine that eases the suffering. Letters from New Zealand and England endorsed this thinking. One woman, for example, wrote that she had suffered from CFS for several years and that she experienced an easing of the usual symptoms after drinking some bottles of Kombucha.

Mrs J.E.B. from New South Wales wrote to me: *"I have been taking Kombucha Tea for two months as I have been ill for four years with ME/CFS. When I spoke to you I had been free of one of the many symptoms, namely headache and fogged brain for five weeks, and there are signs of overcoming the type of Irritable Bowel syndrome that appears with this illness. However ME/CF Syndrome is characterised by periods of remission and relapses, with relapses becoming shorter, hopefully, and less severe as recovery progresses."*

Mrs S.H. wrote *"I have had CF Syndrome for some years and have tried many different products i.e., vitamins etc. to no avail, I have found Kombucha products very helpful."*

My observation with CFS is that people suffering from this disease sleep in bedrooms with high earthray and electrosmog pollution and one cannot say that this is merely a coincidence. Electric measurements within the bedroom would, therefore, be strongly recommended.

RHEUMATISM

When compared to Interferon, Dr Wiesner found Kombucha was approx. 92% as effective in the treatment of rheumatism. One of his patients, however, reported that, after having been treated unsuccessfully with other products, she found, upon trying Kombucha, that her fingers were without pain and her normally painful arm could be moved easily. After further taking Kombucha, she was totally free of rheumatism.

ARTHRITIS

There have been many reports of healing and improvement with arthritis. However, it is not clear how far advanced the state of the illness had been with individual patients. It is certainly important that arthritis should be treated as early as possible and that changes in the diet of the sufferer should also be considered.

The acidity or pH level in Kombucha is of great importance. Questions about its pH level are frequent. The interaction of the pH in the food we eat and how it affects the pH in our body is very complex. The pH scale goes from 0 (acid) to 14 (alkaline). Healthy unpolluted rain-water has pH neutral 7. Lemon juice lies at pH 2; strawberries at 3·3; raspberries at 3·4; potatoes at 5·8 and salt about 7·5. Fermented Kombucha when it tastes like cider has a pH reading of around 3. With varying tea/sugar quantities and fermentation time, the pH can vary between 2·8 and 4. With newfound pH-awareness, many people have become unsure about what is or isn't acceptable and drawn wrong conclusions.

A woman from South Australia wrote to me that Kombucha could not be good for arthritis since it had a pH reading of approximately 3. She is not alone in assuming that Kombucha must have negative effects on arthritis. Practical experience and the fact that a 'sour' or acid food does not necessarily create an acid milieu in the body, show that this conclusion is wrong. Another person, worrying about the pH levels in the body bought Litmus paper and examined the pH of his urine every day to establish whether the pH in his body was correct. His diet was adjusted accordingly to achieve a pH reading of the urine of 7·4. One can imagine that this person gave up after a short time. The reading probably indicated that his body was functioning as it should; flushing out the excess acid.

When someone talks about an acid body, what does it mean? There are many liquids in our body with quite different and specific pH levels. In the digestion, for example, saliva has a pH 7·1; stomach-juice a pH 1·5 and the liquid of the pancreas a pH 8·8. *(Aihara)*. Despite these variations in the body, the blood has very constant pH reading of 7·4. Young people have a lower reading; with increasing age the pH of the blood becomes more alkaline. Fluctuations of the pH in the blood could lead to severe disturbances. A blood pH of 7·7 can result in tetanic convulsions

and a blood pH of 6·95 *(Fasching)* pH 6·0) could lead to uncon-
sciousness and death *(Aihara)*. A pH level over 7·56 sets the stage
for the development of tumours *(Fasching)*. It might be confusing,
but acid food (pH less than 7) has an alkaline effect in the body.

Mrs.T. from I. wrote: *"Kombucha has helped my arthritis,
my writing hand is now no longer in pain."*

Mr. W.R. from M. writes: *"In April 1993 I went on a trip to
Tibet from which I returned after suffering a severe bout of altitude
sickness. About three months later I contracted rheumatoid arthritis. I was eventually placed in the care of a rheumatologist who
put me on a severe pill program to control my problem. In June/
July 1994 I was introduced to Kombucha which I have been taking
ever since. On my visit to the specialist in December, after blood
tests he said I was clear of arthritis. For the last three months I have
not taken any medicine other than two tablets each evening. Pri-
mrose oil once a day and I feel great."*

Another enthusiast wrote: *"Since drinking the Kombucha tea
in December 1994 my osteo-arthritis has improved 50% at least and
I can enjoy life again. Many times I thought of ending it due to
pain. Diverticulitis, ear infections, bowel problems have all improved greatly, I can exercise now and my joints don't swell."*

GOUT

Acute gout is only the tip of the iceberg or, as one naturopath has
said, 'the result of many food mistakes'. Intestinal disorders can
be genetic, which means that people who have this disease in the
family should, by all means, avoid foods which lead to a high level
of uric acid. Pain is caused by the build-up of many small crystals
in this form of uric acid. In early days, gout was called the illness
of the rich. It can be traced back to a far too rich diet. In the so-
called starvation years after the World Wars, gout was one of the
rarest of illnesses. Coffee, however, is often named as one cause
of gout.

Dr Helmut Golz has written that alcohol contains the component purin which is transformed in the liver into uric acid. Half a gram of uric acid is normally discharged daily by a healthy human being. Uric acid, however, is increased by alcohol intake, which leads to gout. Black tea, also known as Russian tea, has the highest level of purin of any kind of tea, according to Dr Golz. Since only this type of tea was used in Russia and Germany for trials, I assumed that Kombucha would have a negative effect on gout sufferers. However, Dr Golz has explained that the Kombucha symbiosis requires purin for its own digestion, during which uric acid, which is generally hard to dissolve, is turned into an aqueous solution and is easily discharged from the body via the bladder. The Kombucha fungus thus transforms both sugar and purin components during digestion, with the result that they are not harmful to the human body.

KIDNEY DISORDERS

In his comparative trials with Interferon, Dr A Wiesner found that the biological food Kombucha was 89% as effective. Dr Harnisch reported the case of a sixty year old public servant whose kidneys had not functioned well since childhood. His uric acid level was also too high. No other remedies (including alternative remedies) helped until Kombucha was taken. His kidneys have worked properly ever since.

CHRONIC COLDS

It is said that a cold generally lasts about fourteen days when one goes to the doctor and two weeks if treated with favourite remedies at home! In other words, there is absolutely no difference. No one remedy has yet been found to alleviate the uncomfortable effects of influenza. Constantly, new and more resistant viruses make the fight against influenza harder and harder. Even Kombucha users are not resistant to flu, although they claim they rarely become ill. Chronic colds, however, have a different cause and should be examined in a little more detail. Often the cause comes from an unhealthy diet – i.e. from a non-healthy digestion. In these cases Kombucha, which has the benefit of being a natural antibiotic, is a recommended treatment.

KOMBUCHA AS A REMEDY FOR TONSILLITIS

In the Russian clinic of Omsk, people suffering from inflammation of the tonsils were treated with Kombucha tea. They gargled with it ten times a day, keeping the fluid in the mouth for up to fifteen minutes each time. The inflammation of the tonsils was reduced and, as with other illnesses where Kombucha was administered, positive side effects could be observed in the healing of intestinal diseases and nasal cavity inflammations.

BRONCHITIS

An improvement with bronchitis is often possible through the drinking of Kombucha. A doctor from the Netherlands *(Harnisch)* said that, after experiencing the benefits of Kombucha for himself, he successfully prescribed it to children suffering from bronchitis.

CANDIDA ALBICANS

Some people suffering from candida albicans only have to think about yeast to have a negative reaction! In an interview with *Search for Health*, Kombucha expert Günther Frank mentions that the Kombucha yeast components help in the fight against candida instead of aggravating it. Split yeasts *(Schizosaccharomyces)* do not contain the spores from which so many people suffer. On the contrary, they can be helpful against damaging yeasts.

It is claimed that the yeasts found in Kombucha will over-run other yeasts, but need time to do so. Candida sufferers should drink the beverage only in moderation and should be careful not to overstrain yeast sediments when bottling the tea. This also applies to non-candida use.

PERNICIOUS ANAEMIA

"After I received a Kombucha fungus from a friend in the beginning of January I have been taking the tea every day. Since 1977 I have had pernicious anaemia requiring monthly injections of vitamin B12. Before this condition was detected my legs started from the foot soles up to get numb, till well over the knees. After a couple of injections of B12 my legs went back to normal except for my foot soles – they remained feeling like 'wood' – no feeling in them at all. For 18 years doctor said I should not expect any improvement as the nerve ends were probably too damaged so I forgot about it as well as a sort of rash around my ankles – some allergic reaction to the injections.

"Now this week I woke up one night because of a stinging pain in my toes and in the morning when I got up I suddenly noticed that I could feel the structure of the carpet under my feet, very remarkable as I never felt that before. And now during the week my foot soles went back to normal and no more numb feeling at all. Also the rash has disappeared."

BEAUTY THROUGH KOMBUCHA

The first time I heard miraculous claims being made for Kombucha as a beauty treatment I was very sceptical. Kombucha as a cure for wrinkles was something I regarded as being too far-fetched. Once wrinkles are present, or so I assumed, it is impossible to iron them out with Kombucha. I did think, however, that it might be possible to use it to prevent them forming. As a side effect to Kombucha's help with the digestive system, it should, in theory, be possible to improve unhealthy, rough skin, which often comes from a one-sided, unhealthy diet or a bad digestion.

I don't believe that drinking Kombucha will banish wrinkles. But don't give up; you can also apply Kombucha externally in the form of an ointment. It is not hard to make. The fungus itself can be pulverised in a few seconds into a cream using a conventional kitchen mixer. Apply the cream to your face, leaving it for twenty minutes while you lie down and rest. Then rinse off with warm water. The cream, with its large content of living yeast components, is also an effective moisturiser. If you consider that the active ingredients of most expensive skin-products are preserved yeasts, you can assume that active and living yeasts should have even better success. With cellulitis, a second treatment, involving massaging the affected areas of skin, daily, with Kombucha tea, can be tried.

KOMBUCHA AS A POULTICE OR COMPRESS

The author received the following letter:

"Noticed your book on Kombucha at the Organic Oasis recently at Forest Glen, and your request for feed-back.

A friend gave me a starter mushroom some months back and we have been using it ourselves and sharing it with others. I'd like to tell you about a sixty-six year old lady who has used the Kombucha in rather unusual ways with incredible results.

Margaret has been under a great deal of stress for over two years since her husband left her, and this has affected her health severely in a number of ways. She has had constant pain in her left hip and lower leg for many months, and when we arrived to stay with her for a short time recently, we found her unable to lie down, and was regularly sitting up in a chair at night to try to get some sleep. For her to lie down in bed was unbearable. She also suffers with hypoglycemia and was afraid to take the Kombucha internally, so decided to try the mushroom as a poultice on her hip. She covered the membrane with "glad-wrap" [seran wrap/cling film] and some old panty-hose, and within only minutes, the pain began to subside. She gingerly took herself off to bed about 9 p.m. and actually slept right through to 4 p.m. She was so delighted with the pain relief that she now has Kombucha mushrooms growing in many containers and is continuing to apply the poultice whenever the pain recurs.

She also suffered from constipation and so one evening before retiring, applied a mushroom to her abdomen. This also proved extremely effective and her bowel movements have returned to normal.

Margaret also has very fair skin and had a nasty growth on the side of her face in front of her left ear. A few Kombucha applications and this growth has recently reduced in size and is almost gone.

This dear lady is no foreigner to natural healing methods, as she is President of a Natural Therapy-Group, and has tried many other alternative medical treatments. None has proven so helpful as Kombucha.

This very clearly has shown me that this special ferment has incredible potential to help alleviate pain and disease in suffering humanity.

I trust you'll find this interesting and helpful".

"My husband suffers from rheumatoid arthritis in his right knee. He has had several operations in the past five years. Despite this his knee is still hurting him, swelling up from time to time and causing great pain. Some days he can't walk or sleep because of this pain. My friend gave me some of the Kombucha fungus she received recently to try it for my husband's knee. I applied the fungus directly to the affected knee as a compress. For the first few minutes he felt a sharp pain right across the joint, but the pain gradually dissipated. It had a soothing cooling effect right away. He was at ease and without pain. We left the compress on the knee overnight to see what effect it had. He slept in peace without pain for the first time in months. We will continue the therapy to see if it really will help this debilitating sickness."

THE KOMBUCHA BATH

Pastor Kneipp, one of the most famous natural healers in the German speaking countries, has developed a wide variety of water therapies, which, today, are being used in six thousand hospitals and private clinics. The human body can easily absorb healing substances through the skin. When added to a bath, Kombucha has good results in energising the whole body and improving general wellbeing. As a general beneficial treatment for common skin problems, add approximately half a litre of Kombucha tea to an average bath. For the treatment of shingles, eczema and skin fungi, Kombucha compresses are recommended. Packs soaked in Kombucha and applied to the scalp can also help to develop stronger and more beautiful hair.

SLIMMING WITH KOMBUCHA

When talking about slimming, we first have to establish what exactly we mean by slim. The desire to be slim has produced such a 'fat market' for industry, that it is very difficult for the average, slightly overweight person to make the right decision. In 1993, 65.6 million Deutschmarks were invested in advertising by the 'slimming' industry *('Wirschaftswoche', Hamburg, Germany, issue 12/1993)*. Similar huge figures were spent in the USA, the UK and most other image-conscious western countries where diet, fitness and the marketing of special, calorie-reduced products are big, big business. There is a personal ideal weight for everyone who is reasonably healthy and realistic about their physical image. But many people, especially women, make their lives miserable by setting themselves quite unrealistic goal weights and constantly changing their diets.

New 'miracle' diets are constantly being offered in newspapers and women's magazines. Those who slavishly follow these badly-balanced eating regimes run the real risk of making themselves ill through malnutrition! There are no miracle treatments that can, for example, give an average-sized woman the willowy frame and body structure of a model. Before embarking on any weight-reducing programme, you must ask yourself whether you prefer to be slim and sick or healthy at your personal ideal weight. Being too slim is just as dangerous as being grossly overweight. As with many things in life, the aim should be to find a happy medium.

So what has all this got to do with Kombucha? In the long-term, only a sensible diet can lead to the maintenance of good health and an ideal weight. But Kombucha can can have a beneficial effect. For use as part of a sensible slimming programme, the tea should be fermented for up to fourteen days, by which time it will have acquired a sour taste. A glass of the beverage should then be drunk early afternoons and evenings, prior to meals. Once the required weight has been achieved, further brews if the beverage should be fermented for only six to ten days and drunk as part of a balanced diet. Kombucha supports the digestion processes and, therefore, helps to achieve an ideal individual weight. It has a mild purgative effect which is more noticieable when taken before meals. In contrast to laxatives, however, Kombucha does not increase its function with larger dosages.

In an age in which many people do not have enough exercise, have sedentary occupations, eat irregular meals and include much processed food in their diets, Kombucha is especially recommended. A sluggish bowel will lead to constipation, excess weight and many other symptoms associated with an unhealthy lifestyle. The positive effect of Kombucha on health can definitely be related to the effect it has on the regeneration of intestinal flora. Scientists world-wide agree on that point.

When I released my video in 1984, 'Back to nature with medicinal herbs', in which I included references to slimming and ideal weights, one famous women's magazine took notice. Following an interview with me, the magazine used the method I recommended in a trial which was successful with every person taking part. However, as I refused to change my report and turn it into a diet plan, the story was never published. Since that time, this magazine and many others have published countless diets – each time with corresponding increases in their sales. However, while any slimming programme however badly-balanced nutritionally and deficient in fresh food, may, in the short-term, lead to weight loss, the long-term results are bound to be ill-health. Health depends on balance. An inadequate diet will never be successful in healing illness or achieving a sensible, permanent weight loss.

KOMBUCHA FOR THE YOUNG OVER 80s

Any report claiming that a 130 year old man and an 89 year old woman were able to have a healthy baby was bound to put the media in a frenzy and draw attention to Kombucha. We can each form our own opinion as to the accuracy of the ages given. However, what should be noted is that many older people believe that Kombucha is the reason for their still-active life. Even Dr Sklenar, who was a military doctor in Russia, reports coming across cases of people of advanced years who appeared to be leading healthy, energetic lives. Kombucha has the effect of de-toxifying, aiding the digestion and improving energy. It is not surprising, therefore, that it contributes to the health and fitness of people of advanced years. Kombucha can only be regarded as part of this successful recipe, however, and should not be re-garded as its sole cause.

KOMBUCHA AS REMEDY FOR IMPOTENCE

The potency of a human being is widely-regarded as an accurate measure of health and vitality. In 1992, Germans spent 790 mil-lion Deutschmark on sex hormones *(Stern 12/1992)*. That is al-most 10 DM per head of population - from new-born infants to great-grandparents. Stress, unhealthy diet, environmental factors, lack of exercise and side effects from medicine can all contribute to a decrease in sexual ability. In his research in Russia, Dr Sklenar found many examples of older people who were still very active sexually. The old farmers in the rural regions he visited attributed this ability to drinking Kombucha. According to the

German naturopath Jost Kuessner *(das neue)*: 'The fungus helps impotence. Many men when they are still young have problems getting an erection. After only one week, Kombucha will help sexual vitality.'

WOUNDS AND ULCERS

The fungus itself can also be used to heal wounds and ulcers. A woman from Queensland told me that she suffered for many years from a leg ulcer. She bandaged a fresh fungus direct from its brewing container on to the ulcer. The ulcer healed in less than a week and did not return. Other people have also reported good results from putting a young fungus directly on to wounds, following injuries.

'DOPING' WITH KOMBUCHA

Members of one of the largest 'sporting' organisations in the world, the Russian military, regularly drink Kombucha. Particular regiments have their own secret recipes for the beverage and the German army has also conducted trials on it effectiveness. Professor Dr Simon Gerrit of the Army Sport School in Warendorf, Germany, tested Kombucha and came to the conclusion that this "pure biological Kombucha fermented tea" had a "strengthening effect and improves the performances of the athletes".

Russian high-performance athletes are given Kombucha to increase their performance. Scientific tests were conducted in the Olympic centre in Warendorf, with exceptional results. Even with the hardest training there, muscular aches were no problem and the athletes improved their performances through drinking 1 cup US (0.2 litre) of Kombucha three times a day. The team of doctors conducting the trials found that salt in the lactic acid *(Lacta readings)* lessened with Kombucha. With high bodily exertion, the organism sours more quickly and Kombucha counteracts this. Overall the athletes were positively affected and recovered

much quicker after long runs. Increased wellbeing and performance, according to the scientists, were due to the positive ex-
change of energy in the cells, encouraged by the Kombucha.

One man asked me many questions about Kombucha. He
mentioned that he was making 30 US pints (15 litres) of Kombucha every day, for his own consumption. I calculated that he
must have a family of thirty, based on average Kombucha consumption! But no, it turned out that he had been giving it to his
racehorses as an 'energy shot'. 'I know that the rewards of Kombucha brewing are 'measurable' in valuable seconds in a race', he
said. There are also reports that racing camels in Arabic countries
receive a fermented drink as a form of legal 'doping', as the Swiss
health magazine *Natürlich* reports under the heading 'Bio-Strath
makes camels fast'. Of the production in Switzerland, 73% is
exported. The ingredients and details of the fermentation are the
manufacturers' secret. Many Australian herbalists and
naturopaths have spent years ministering to their patients with a
secret preparation - which turned out to be nothing other than
Kombucha.

KOMBUCHA USED IN VETERINARY MEDICINE

Food growers have received more and more criticism in recent
years because of their dependence on chemicals. Today the call
for natural food is becoming stronger and stronger. Biodynamic
produce and alternative healing methods for animals are slowly
becoming popular. Kombucha is also being used for veterinary
applications. In a trial with sheep and calves, Kombucha drops
were applied to animals suffering from diarrhoea with a 100%
success rate. With healthy animals, Kombucha was mixed into
their food, with a growth increase of 15%.

The economics of Kombucha-drinking are, of course, also
important. Kombucha fermentation requires a large amount of
sugar. Would this not be a beneficial use for the excess sugar we
consume and which accounts for a lot of our illnesses? Would it

not also be possible to replace health-endangering hormone treatments on animals with a healthy Kombucha product?

A delightful anecdote from an Australian correspondent:
*"The most startling success story for me has been with our family
pet, a German shepherd dog, who has suffered for the past few
summers with an itchy eczema, resulting in a bald patch near his
tail – this year some warty type lumps appeared. As it is hard to give
him the liquid I began to cut the fungus and give him a piece each
day, he now looks forward to his piece of Kombucha, before
breakfast each day. The hair on his back is almost fully repaired
and I expect the warts will also disappear within time. I will continue using the Kombucha until they have disappeared. He also
seems to be a whole lot stronger – his back legs which were
becoming weak."*

NETWORKING KOMBUCHA

The way that the AIDS community in California has discovered a
sense of compassionate sharing of the fungus is a wonderful example of the networking magic of Kombucha. There is probably no
other organism that reproduces itself so prolifically every week. It
seems wasteful to throw it away when it could bring healing to a
friend, or to some stranger who could become your friend. Just as
Mother Nature has given us this priceless gift of healing, you can
pass it on to someone else.

An excellent way to network Kombucha is to keep a record
of those to whom you give cultures. Then, when others ask you
for one and you have not one to spare, you can simply refer them
to someone on your list. Try putting up a notice in your health
food shop. Just imagine how our society could be transformed if
people got into the habit of growing a culture in their kitchen, just
like they used to in deepest Russia!

You don't have to be a mathematician to work out that, if
everyone you gave one to in a year did the same, there would be
thousands of Kombucha cultures out there helping to heal other

people by the end of the year! If you all then shared recipes* and exchanged tips on ways of making the brew more interesting or of turning it into ointments, what a community builder this would be. Just what we need!

The word about Kombucha is spreading like wildfire. Thousands of people are cultivating cultures and passing them on. If you want to start one, keep your ears open, look in alternative health magazines or ask holistic therapists. Please don't write to the publishers for a culture. There are some contact addresses on p.106 for those who would like to register as a source of cultures in their neighbourhood or want to obtain one.

*Look out for the Kombucha recipe book, to be published by Gateway Books.

EPILOGUE

HEALTH IS BALANCE

Without exception, all Kombucha researchers recommend it as an overall healing method. One cannot isolate certain illnesses and their treatment if holistic health is to be restored. In recent years, there has been a significant resurgence of interest in this healing method. Kombucha should not, however, be regarded as an isolated healing remedy, but rather as what it really is, a very precious, living and health-giving food. If one asks the elderly to what they attribute their good health and fitness, the usual answer is a quizzical look. If one presses further, however, and asks specific questions, the answer that often emerges is, 'to a healthy, balanced diet'.

NEW AGE – OLD AND NEW

Countless 'alternative' healing methods can be found all over the world, but some can become muddled in their migration to other countries. Releasing energy blockages with Reiki, experiencing aura diagnosis using Kirlian photography or plasma print; unlocking fears with 'past-life regression'; mastering the present and future with numerology or astrology; healing with sound or colour, aromatherapy, acupuncture, crystals - these are a few examples of many healing practices. Many of them represent the rediscovery of traditions which go back thousands of years. Better communication has spread this knowledge rapidly. No single tradition or practice should be dismissed, but needs to be considered among the multitude of healing methods available, including modern medicine.

PLACEBO – MODERN MEDICINE'S DISCOVERY

The phenomenon of the placebo effect does not fit into the clear picture of natural science. If modern medicine were to use words from the general vocabulary, they would, perhaps, describe it as the 'miracle effect'. Healing with **nothing** is scientifically imposs- ible. Healing with a word or a touch is something we are only familiar with from the Bible. During extensive testing of modern medicine, however, we frequently come across the self-healing or placebo effect. Did science rediscover miracle healing in these trials and did they prove it? Even though placebo medicine does not have any active ingredients, it has remarkably high healing successes for all types of illnesses. Even in 'double blind' testing, the placebo effect has been identified.

For approximately fifty years, research has been testing the effectiveness of placebo medicine. In tests, one group of patients is given a new medicine while, at the same time, a comparative group is given the same product, without any active ingredients. In double blind testing, neither the patient, nor the doctor know which drug is being taken. If, for example, two thirds of patients, who suffer from strong headaches respond to the real drug while the same number of patients in a second group react identically to the 'pretend' drug, this result would not lead to the conclusion that the active component was functioning. Everyone wants to have a local anaesthetic at the dentist. How would a patient react, however, if, after treatment the dentist revealed that the 'ana- esthetic' used was only water? About 30% of patients show the same result with a placebo as they do with an actual drug.

The healing effect with a placebo depends to a great extent upon the person administering it. A confident administrator gets a better result than one less self-assured. It is interesting to note that negative side effects from using the the real medicine are also experienced with the placebo medicine. Evidently, the brain is capable of giving healing messages to the body. When we receive medicine, we believe in its success in healing and, therefore, subconsciously heal ourselves. If the self-healing potential of the mind is so powerful, should we not practise this power in order to heal ourselves without any negative side effects? After I published the first edition of this book, I received many letters and telephone calls. The feedback in the context of placebo medicines was particularly strong.

A nurse who, for many years, worked on hospital night shifts told me that it was common practice to give patients a placebo. Patients first received a regular dose of medication to help them sleep but many later asked for a second pill in the middle of the night. The maximum dose permitted per night was one tablet, so the nurses gave those patients requesting a second dose a vitamin C pill (visually identical to the sleeping pill) instead - with a 100% success rate! When I questioned this high success rate, on the grounds that vitamin C has a stimulating effect that is exactly opposite to the result one would expect from a sleeping pill, the nurse only replied, "It simply worked, that's it". My question as to why the safer vitamin C had not, therefore, been given to the patients in the first place was not answered. I can only repeat my question: Why are the millions of dollars which are spent every year on researching new pharmaceuticals not used in the same proportion for researching the possibilities of increasing the 'placebo effect' so that fewer pharmaceuticals would be needed? "When you take a drug with a glass of water to swallow it, the water is often doing you more good than the medication."
[F. Batmanghelidj, MD]

FASTING FOR HEALTH INSTEAD OF SUICIDE WITH KNIFE AND FORK

Dietary problems are occurring as never before. We see on our television screens that countless people are daily starving to death in many countries while, on the other side of the world, many people are suffering health problems from eating too much. The question now is, which death is preferable - rapid starvation or slow death through overeating? It may sound strange, but fasting must be regarded as one of the better healing methods. The positive health effect of fasting was recognised by many early religions. One complete day of fasting, or a time of fasting over some weeks with only limited food, is found in most religions. Animals also fast themselves healthy, i.e. they fast during long periods of exertion. The eel and the salmon both fast when travelling long distances to their mating grounds. Migrant birds travel thousands of kilometres without taking any food. Fasting is a highly effective tool towards preventing illness.

Intestinal illness, excessive blood fat and high blood pressure are the main risk factors in the Western world, all of which could easily be treated with the problems of the Third World countries. If the Western world were to supply the food saved from one day of fasting to Third World countries, there would be no people dying from over-eating on the one side and no people dying of starvation on the other.

ACID AND ALKALINE FOODS

There are two groups of acid and alkaline foods. One is made up of foods that are actually acid or alkaline and the other of foods which produce an acid or alkaline reaction. For example, lime is extremely acid with a pH reading of 1·9, but this fruit increases the alkaline content of the body. If you want to influence your body's pH, it is not so important to know the pH of any particular food you eat, but to know what reaction that food will have in your body. For more information, I recommended a study of the book: *Acid and Alkaline* by Herman Aihara.

Research from Russia and Czechoslovakia is interesting in this context. Water, for the purpose of this research, is divided into 'live water' and 'dead water'. DC current converts neutral water of about pH7 into acid water (pH 4) and alkaline water (pH 10). This water was used with excellent results, according to reports, with a wide range of diseases. The research was based on the fact that plants can thrive only in a specific pH milieu. Unwanted stinging nettles for example can be eliminated in two ways. One is to spray with chemicals, with all their negative environmental side effects. The other is to alter the pH of the soil (calcium) so that the stinging nettles can no longer thrive.

There is a similarity with diseases, according to the research. Altering the pH of the blood should fight certain diseases. With cancer for example, if the blood pH is altered with activated water to 7·4 there is no 'fertile soil' for the cancer to grow in, which should prevent further growth of the tumour. As the information from Russia shows, over 500 patients with different diseases were all treated with success. It is also pointed out, however, that it is impossible to cure all diseases with this water. *(Krotov)* The manufacture of this water is relatively easy. I have met in the last months some people from Czechoslovakia who confirmed with enthusiasm the healing successes achieved with activated water. You can find detailed information about 'live water' and 'dead water' in my book about water-cures.

TESTIMONIALS

Thank you everyone who wrote me about their experiences of drinking Kombucha. The following is a sample of these:

This is from an Australian doctor: .... *I have only been using Kombucha for the last 6 months, and have a quite a few patients trying it out. So far the main results coming to hand are as follows: 1. Energy increase. 2. Weight loss. 3. Regular bowel motions. 4. One patient with outstanding stomach problems is now 100%. 5. Moles diminishing in size. 6. Good results with sinus and mucus elimination. 7. I used it during the hay fever season and found relief much earlier than usual. The only problem struck so far is with hives if taking too much. We now have fifty families taking Kombucha.*
And a remarkable testimony from H.T. of S: .... *My father was diagnosed with lung cancer. He had cancer in one lung the size of a*

small orange. His doctor advised him that the cancer was incurable but treatment would give him a better quality of life for the time he had left. His brother gave him a Kombucha which he started drinking immediately – 2 glasses a day. Four months later an X-ray showed no trace of cancer. We the family have no doubt the Kombucha was responsible for his recovery.

.... Have recently been introduced to the miraculous Kombucha. I am ecstatic with the results to date regarding constipation and stomach pain which no longer haunt my days.

.... I must say that Kombucha picks me up. I have asthma from working in a clothing factory for years. I have had chronic fatigue for many years and, after a couple of bottles of Kombucha, I felt a marked difference.

...... I have had chronic fatigue for some years and have tried many different products i.e. vitamins etc, to no avail. I have found Kombucha products very helpful.

...... It has become quite 'a craze' at the high school here among the teenage girls who are growing fast and always tired, and they say it has given them more energy – many thanks.

...... I have been buying this medicine (Kombucha) from a naturopath for eighteen months and it has saved both my husband and myself from dying.

.... Kombucha has helped my arthritis, my writing hand is now pain-free.

.... I thought you would like to know that people with high blood pressure may have to reduce medication while taking Kombucha, as a lady's blood pressure went too low.

.... For myself, I am greatly impressed with this tea (Kombucha). I am not young, 77. My digestion is much better and I have far more energy. A friend of mine, a couple of years younger than myself, has been able to cut her blood-pressure tablets by half. The side effects of the tablets have eased. A highly irritating rash all over her body is now disappearing. She has also effectively rid herself of a clot in her leg by applying tea compresses for three nights. She has always been subject to these clots. They can last three or four weeks and be the cause of much pain. If I hadn't seen your book in a

health shop here in M., well these positive results would not have come about. So thank you for writing the book & acquainting us all about this wonderful Kombucha tea.

.... *Benefit greatly from Kombucha. Even after a week I'm feeling much beter – regular bowel movements, more energy, sleeping well;*

.... *I have been drinking the tea for about a month and have noticed a change in my varicose veins on my right leg;*

.... *I have been brewing and drinking it for three months and feel more energy and am generally in good health.*

Some typical comments about the varieties of taste experienced:

.... *delicious;* ...... *my children thought it was apple juice;* .... *I enjoy Kombucha since it is the only drink I like without negative effects;* .... *I feel really well after drinking Kombucha;* .... *since I started drinking Kombucha I like it more then beer or wine and I feel much better the next day;* .... *I have brewed Kombucha for a year and am still finding new delicious recipes;* .... *really healthy and, at the same time, delicious.*

Some more comments about the effects after drinking Kombucha:

.... *cancer growth in the lung shrunk;* .... *a 79 years young man has taken tea for two months, has circulation back in his hands, they were black and blue, and are now a much healthier colour;* .... *he had a tendency to retain fluid, now he is slim with trim ankles;* .... *noticed grey hairs on chest and tummy turning black again;* .... *my blood pressure went back to normal;* .... *brown spots on skin disappearing;* .... *urine became clear, from being very cloudy;* .... *better sleep within two weeks;* .... *pain relief from arthritis in shoulders and neck;* .... *my hair thickened up considerably, with less loss when shampooing;* .... *constipation was a problem for twenty years, with Kombucha everything is just fine;* .... *urine cleared, weight easier to maintain;* .... *I simply feel so much better;* .... *increased energy and mobility playing tennis;* .... *bowel regularity,* .... *spots from the sun disappearing;* .... *hot flushes disappeared;* ....

APPENDICES

Appendix One: BIBLIOGRAPHY

Aihara, Herman, 'Acid and Alkaline, George Oshawa Macrobiotic Fndtn, 1544 Oak St, Oroville, CA: (1980).

Biser, Sam. Interview with F.Batmanghelidj, MD 'The Greatest Health Discovery in the World'.

Breuss, R. 'Cancer and Leukemia - Advice for the prevention and natural treatment of numerous diseases': Bludenz Vgb, Austria, (1982).

Brown,A.J. 'On an acetic ferment which forms cellulose': *Jour.Chem.Soc.Lond.* (1886)

Carstens, Dr V. 'Hilfe aus der Natur - meine Mittel gegen Krebs': *Quick*(43/1987)

Cribb, A.B. & J.W. *Wild medicine in Australia*: p.60 (1986)

Das Beste, 'Geheimnisse und Heilkraefte der Pflanzen'.

Filho, L.X., Paulo, E.C., Pareira E.C. & Vincente, C. 'Phenolics from tea fungus analysed by performance liquid chromatography': Phyton, Buenos Aires (1985)

Frank, Günther, W. *Kombucha - Healthy beverage and natural remedy from the Far East*: Ennsthaler, Steyr, Austria

Gadd, C.H. 'Tea Cider': *Tea Quarterly* (Talawalelle, Ceylon, 1933)

Gold Coast Bulletin: 'Cancer plea bears fruit' and 'Pawpaw's medicinal qualities': Weekend Bulletin (1993), P.O.Box 1, Southport, Qld 4215

Golz, Dr. *Kombucha Ein altes Teeheilmittel schenkt neue Gesundheit*, Ariston, München, Germany,)1992).

Goetz, Georg 'Kombucha - der Wunderpilz, der Millionen Gesundheit schenkt' in *Das Neue* (Issue 3 - 14, 1988)

Harnisch, Dr Guenther: *Kombucha geballte Heilkraft der Natur'*,Turm, Bietigheim-Bissengen, Germany.

Harris, R.D. *Prostawort or Willowherb*: Candelo, Crowsnest, Australia

Hess, Walter, 'Bio-Strath macht auch Kamele schneller' in *Natürlich*, AT Zeitschriftenverlag, Bahnhofstrasse 39-43, Aarau, Switzerland (1994)

Horstkorte, C. 'Zaubertrank aus China-Pilz hilft auch bei Sex Problemen' Kaminski, A. 'Aerzte: Pilz heilt Frauenleiden': *Bild der Frau* (2/1988) a. Springer Verlag, Hamburg, Germany.

Kaminski, A. A. 'Aerzte: Pilz heilt Frauenleiden': *Bild der Frau* (2/1988) Springer, Hamburg, Germany.

Kanuka-Fuchs, Reinhard. Building Biology & Ecology Institute of New Zealand, 22 Customs St.W., Auckland, New Zealand

Kelly, Justin. 'Magnetic Health Teaching Services': 5 Cocararra Ct, Tugun Heights, Qld 4224. Australia.

Koerner, H., 'Der Teepilz Kombucha': *Der Naturarzt* 108 (1987), and 'Kombucha - Zubereitung wurde von Sportmedizinern getestet' *Natura-med* (10/1989)

Kuski, A. & Esko, W., Avery Publishing Group, Garden City Park, New York, USA

Lassak, E.V. & McCarthy T., *Australian Medicinal Plants,* Methuen, 44 Waterloo Rd, Norgh Ryde 2113, Australia (1983).

Meixner, Dr A., *Pilze selber zuechten*: AT Verlag, Aarau, Switzerland

Mulder D., 'A Revival of Tea Cider': *Tea Quarterly*, Talawakelle, Ceylon, (1961)

Perko J., 'Kombucha - Health you can drink' and other information. The Way To Live Sanctuary, Canungra, Qld 4275, Australia

Potter's *New Cyclopaedia of Botanical Drugs and Preparations*

Quinn, D., *Left for Dead*: Quinn Publishing Co., Minneapolis, U.S.A.

Reiss, J., 'Der Teepilz und seine Stoffwechselprodukte': *Deutsche Lebendsmittelrundschau* (9/1987)

Search for Health: 'Kombucha 'yeasts' fights candida, they do not encourage it'; and 'Kombucha converts tea and sugar into a healthy, nutritious detoxifying beverage'.

Sharma, Dr P.R., *The Art of Spiritual Living*: The Ram-Rukimini Institute Liaison, Geneva, Switzerland

Teeguarten, Ron, *Chinese Tonic Herbs,* Japan Press (1986).

Tietze, Harald, 'Back To Nature With Medicinal Herbs' (Video 1984)

Tietze, Harald, *Earthrays: The Silent Killer*: P.O.Box 34, Bermagui South, NSW 2546, Australia

Timmons, Stuart, 'Fungus Among Us *New Age Jour,* PO Box 53275, Boulder, CO 80321

Urban & Schwarzenberg, *Roche Lexicon Medizin*

Vogel, Dr A., *Der kleine Doktor*: Teufen, Switzerland

Waal, de Dr. M., *Medicinal Herbs in the Bible*: Weiser, York Beach, Maine, USA

Wagner,H., *Gegen jede Krankheit ist ein Kraut gewachsen*: Ruhland Altoetting Weidinger, Herman Josef 'Kombucha - Tee der aus dem Meere kam' and 'Die Kombucharunde' in *Ringelblume*: (1988) 3822 Karlstein, Austria

Weidinger, Hermon Joseph, 'Kombucha – Tee der aus dem Meere Kam' and 'Die Kombucharunde' in 'Ringelbume,' 3822 Karlstein, Austria (1988). Wiesner Laboratories: 'Kombucha nach Dr med Sklenar' (1987) Schwanewede

Willner, Robert E., *The Cancer Solution*: Peltic Pub.Co., 4400 North Federal Hwy., #210, Boca Raton, FL 33431, USA (1994)

Zimmerman, W., 'Wogegen hilf der Kombucha Pilz?' *Fortschritte der Medizin* (12/1989)

Appendix Two: NETWORKING SOURCES

In **Britain**, if you want to register as a source of Kombucha cultures in your community, you may lodge your name as such with: *The Kombucha Network, PO Box 1887, Bath, BA2 8YA* (**Please send s.a. envelope**). If you want to acquire a culture, and have not found a local source, they will also try to help you find one, but please be patient; this network will take some time to set up, and it will work best by local osmosis.

In **USA**, Kombucha culture supplier: Lee Vinocur, PO Box 81, North Palm Springs, CA 92258 (tel: (619) 329 9813).

In **Canada, South Africa**: In the next printing we shall try to give some addresses for obtaining cultures. But Kombucha is networking fast, especially in the US West, so keep your ears open, and check appropriate magazines or alternative health sources. If you wish to be a supplier of cultures or books or wish to share your experiences as a Kombucha user, write to the author (address below) so we can update the information in future printings. Listings are free!

In **Australia**:
This book has information which, at the time of writing, can be regarded as being accurate. New results dealing with Kombucha and other alternative therapies can be supplied as fact sheets. If you have experiences with Kombucha or have further information please send them to the address below.

Fact sheets are available on the following themes:
Healthy Kombucha cultures. Commercial Kombucha beverages and test results. Herbal and Green teas – biologically grown and fresh. Temperature control units. Building biological surveyors and additional information. Seminars on: Kombucha brewing. Growing and using herbs. Dowsing. Building biology. Kneipp applications.
Please enclose a self-addressed stamped envelope.
'Kombucha', c/o Harald W.Tietze, P.O.Box 34, Bermagui South NSW 2546, Australia. Tel: 064-934-552 (Internat'l +61-64-934-552) Fax: 064-934-900 (Internat'l +61-64-934-900)

Appendix Three: - OTHER BOOKS AND SOURCES

Kombucha and Pawpaw Leaf Extracts are manufactured by The Way to Life Sanctuary, Lot 7, Climax Court, Canungra, Qld 4275, Australia (Tel: +61.75.435.104: Fax: +61.75.435.263) [It is hoped in the next printing to give addresses of US & UK suppliers.]

Homeopathic Kombucha: Tina White, Manning Natural Healing Centre, 216 Victoria St. Taree, NSW 2430, Australia.

Heating Devices: For Kombucha brewing. Quickheat Industries, 6 Michelle Rd, Christchurch, New Zealand has three heating devices supplied to all countries with a variety of voltages and plugs.

TIETZE PUBLICATIONS, available in Australian health shops or directly from the author: Harald W. Tietze, PO Box 34, Bermagui South, NSW 2546, Australia:–

Water Medicine – by Harald W. Tietze
A practical manual on different water treatments for many health conditions. Includes live and activated water, urine therapy, steam and ice treatment, Kneipp water cure, and understanding water as an energy carrier. Using water to strengthen the body and treat illnesses of home *100pp. $ Aus. 9.80.*

Earthrays: *The Silent Killer* – by Harald Tietze
Explores geopathogenic (geo = earth, pathogen = ill-making) radiation emitted from the earth in certain locations. For thousands of years the harmful effects of earthrays on plants, animals and humans have been observed and understood by many cultures.

Harald's research in Germany, Austria, Italy and Papua New Guinea adds a new dimension to ancient knowledge which raises awareness of the affect of earthrays on modern diseases such as cancer, asthma, arthritis, rheumatism. *72pp. 27 col. illus., $ Aus. 14.80.*

Back to Nature with Medicinal Herbs. *Video by Harald W. Tietze*
Practical hints for growing medicinal herbs, home made health teas; how to make your own tinctures and herbal ointment. Examples for successful self-treatment with medicinal herbs: * Varicose veins * Disorders of

the prostate * Bed wetting * Heartburn * Digestive disorders * How
herbs can help you to stop smoking. Detailed tips about sensible diet to
reach your ideal weight without fasting. Samples to make your home-
made plant protection biologically against ants, greenfly, caterpillars and
other insects. Available in VHS and Beta.

GATEWAY BOOKS publishes works that explore in depth, alternative
world views and life scenarios. Here is a selection to whet your appetite:–

be resolved.
"Everything you ever wanted to know about the Universe, but didn't know who to ask" Kindred Spirit magazine, UK.
360pp, £9.95 $14.95

Cosmic Connections: *Worldwide crop formations and ET contacts* – by Michael Hesemann
The first book to bring together the two most mysterious and unexplained phenomena of our time – crop circles and ETs, showing fascinating connections between them. A global and up-to-date survey of the most convincing information, with many color and black and white photos. The author is a world authority in both these subjects.
160pp, £9.95 $15.95

When the Earth Nearly Died: *Compelling Evidence of World Catastrophe, 9,500BC* – by D.S.Allan & J.B.Delair
Evidence from many disciplines, traditions and cultures, of a cataclysm which nearly destroyed Earth and Mars about 11,500 years ago. The authors draw on decades of research to describe how a golden age disappeared with appalling devastation, and show how their findings could have relevance for present world changes.
384pp, incl. many photos, tables, maps & charts: £12.95 $19.95

A Search for the Historical Jesus: *From Apocryphal, Buddhist, Islamic and Sanscrit Sources* – by Prof. Fida Hassnain
Millions of people have been brought up to believe that Jesus's life mission ended with the crucifixion. Here, a respected Sufi historian finds evidence of information suppressed by the Church that Jesus survived the Cross and undertook an Essene-backed extended ministry in India and the East. Riveting reading.
268pp, incl. many photos & maps: £8.95 $14.95

Living Energies: *About the discoveries of Viktor Schauberger on Natural Energy* – by Callum Coats
Schauberger was a pioneer in working closely with nature. This important book follows the development of his revolutionary ideas on water purification, transport, free energy heating and home power generation. It lays the foundations for a new technology which can save this planet from destruction.
386pp, many illustrations: £12.95 $19.95

Safe as Houses: *Ill health and electro-stress in the Home* – David Cowan &
Rodney Girdlestone
To help you identify sources of electrical stress in the home - from nearby
electrical generators to leaky microwave ovens and circuits; also covers
problems of geopathic stress related to earth energies, and gives instruc-
tion and background to dowsing the energies.
256pp, many illustrations: £7.95 $12.95

Write for our catalogue – we'd be happy to put you on our mailing list:
Gateway Books, The Hollies, Wellow, Bath, BA2 8QJ, UK.
(Tel: 01225-835 127 Fax: 01225-840 012).

In the **USA** Gateway's books are available through many bookstores. In
case of difficulty you may contact our distributors: Atrium Publishers
Group, 3356 Coffey Lane, Santa Rosa, CA 95403 (tel: (707) 542.5400),
or write for our catalogue to our marketing office at 18900 Olive Avenue,
Sonoma, CA 95476 (tel: (707) 939.1953).

In **Canada**, Gateway's distributor is Temeron Books, #210, 1220 Ken-
sington Rd NW, Calgary, Alberta T2N 3P5. (tel: 403-283.0900).
Australia: *"Kombucha" is imported by*: Quest, 484 Kent St, Sydney 2000
(tel: 02.264.7152 fax: 02.283.3772).
New Zealand: Peaceful Living Publications, PO Box 300, Tauranga,
(tel: 7.571.8105).
South Africa: Wizard's Warehouse, PO Box 3340, Cape Town 8000.
(tel: 21-461-9719).
Hong Kong: Pacific Century, 14 Lr Kai Yuen Lane, North Point, Hong
Kong. (tel: 811.5505).
Singapore: Pansing Distribution, 8 New Industrial Rd, Singapore 1953.
(tel: 382-0488).

Index